LINUS PAULING

LINUS PAULING

TED GOERTZEL
AND BEN GOERTZEL

with the assistance of
MILDRED GOERTZEL
VICTOR GOERTZEL

with original drawings by
GWEN GOERTZEL

BasicBooks
A Division of HarperCollins*Publishers*

Published by BasicBooks, A Division of HarperCollins Publishers, Inc.

Designed by Elliott Beard

Library of Congress Cataloging-in-Publication Data
Goertzel, Ted George.
Linus Pauling: a life in science and politics / by Ted Goertzel and Ben Goertzel with the assistance of Mildred Goertzel, Victor Goertzel.
p. cm.
Includes bibliographical references and index.
ISBN 0-465-00672-8
1. Pauling, Linus, 1901–1994. 2. Scientists—United States—Biography.
I. Goertzel, Ben. II. Title.
Q143.P25G64 1995
540'.92—dc20
[B] 95-9005
CIP

---

95 96 97 98 ❖/HC 9 8 7 6 5 4 3 2 1

Contents

ACKNOWLEDGMENTS

My parents, Mildred and Victor Goertzel, began work on this biography, with Pauling's assistance, in 1962, as a continuation of their work on the childhood of eminent people (see *Cradles of Eminence* and *Three Hundred Eminent Personalities*). They admired him especially for his leadership of the peace movement and thought that his biography would be inspirational for young people who might be thinking of a career in science or involvement in worthy causes. They interviewed relatives, teachers, and neighbors who knew Pauling in his childhood. Pauling was generous with his time in reading and editing their drafts of early chapters about his childhood and family background. He also authorized the psychologist Anne Roe to provide them with a copy of the record of the Rorschach test that Pauling had taken as one of the subjects in her book *The Making of a Scientist.* Victor had done his Ph.D. dissertation in psychology using the Rorschach.

Mildred and Victor differed with Pauling, however, about the focus of the book. They wanted to emphasize his personality; he wanted extensive details about his scientific work. They put the project aside for a number of years but continued to follow his life. They were troubled when Pauling began his crusade over vitamin C. Much later, they were distressed by his treatment of Arthur Robinson, whom they came to know personally. They again worked sporadically on the manuscript, which became more critical, less pacifist hagiography.

Mildred and Victor then asked me, their oldest son and a professor at Rutgers University with an interest in the psychosocial roots of political beliefs (see *Turncoats and True Believers*), to help with the manuscript. I interviewed some of Pauling's associates who lived on the East Coast and added material on Pauling's scientific contributions and on the controversies surrounding orthomolecular medicine.

When Pauling contracted prostate cancer in 1991, at the age of ninety, we decided that I should undertake the preparation of a final manuscript. Mildred and Victor's health no longer permitted full participation, and we thought that a complete rewrite by one author would lead to a more coherent book. My experience in statistical research methods was particularly helpful in making sense of the ongoing controversies concerning vitamins and health. Victor and I visited the newly established Linus and Ava Helen Pauling Papers at Oregon State University to collect material from letters and other documents and updated some of the interviews with Pauling's associates. I reestablished contact with Pauling, and he was reviewing the manuscript when he died in August 1994. He had already sent comments and a few minor corrections of chapter 3 concerning his contributions in chemistry.

At this point, I asked my son, Benjamin Goertzel, a mathematician and cognitive scientist, to help with the chapters on Pauling's science. Our editor, Susan Rabiner at Basic Books, told us that Pauling was right: the biography needed much more about Pauling's contributions to science. Ben has the mathematical understanding of quantum theory needed to understand Pauling's work fully and the literary skill needed to communicate it. He also contributed his insights as a cognitive scientist who understands a great deal about how scientists think (his books include *The Structure of Intelligence, The Evolving Mind,* and *Chaotic Logic*). In the end, the final draft covers very much the same scientific material that Pauling told my parents it should thirty years ago.

In no way can this be considered an authorized biography—it is quite critical of Pauling in many ways. Pauling knew of our views because we had published them in the *Antioch Review* in 1980, and we appreciate his willingness to review the manuscript. We regret that he did not have time to comment on the later chapters.

Although this biography was never "authorized" by Linus Pauling, nor did we ever have any agreement with him concerning its publication, we did begin the project with his cooperation and assistance, which we gratefully acknowledge. We have approximately twenty hours of tape-recorded interviews conducted in the mid-1960s with Pauling, and he gave us access to his clipping files, to the diary he kept as a boy, and to other documents in his possession. He also authorized the psychologist Anne Roe to release to us a transcript of the Rorschach and thematic apperception tests, which he took as part of her classic study *The Making of a Scientist.*

For the story of the early years of Linus Pauling's life we are indebted to many persons, including several who are now deceased: his "Aunt Goldie," his mother's older sister, whom we interviewed at length in a nursing home in Portland, Oregon, in 1963; his ninth-grade schoolteacher Grace Geballe; and his uncle, Herman Blanken, who gave many details of the Pauling-Blanken family, who settled in Oswego, Oregon.

We have two letters each from Lloyd Jeffress, a psychologist who was classmate to Pauling during his grade school and high school years and his first year in college, and from Ray Tracy, a Condon, Oregon, writer known for his wild west stories in the *Saturday Evening Post,* who sang at the funeral of Linus's grandfather.

Among relatives who were his contemporaries, we acknowledge the generous assistance of his two sisters, Lucile Jenkins and Pauline Emmett, wife of Paul Emmett, an engineer who was Linus Pauling's classmate in college and at the California Institute of Technology. Linus's cousin, Mervyn Stevenson, son of Goldie Stevenson, made available to us the sixty-thousand-word diary of his and Linus's maternal grandfather, Linus Wilson Darling, a document worthy of publication itself. Another cousin, Ruth Jacobson, gave us explicit information about the Darling family tree. Many of the relatives read early drafts of chapters and wrote back often giving additional information. They gave much more than we were able to use, and we will make their letters and other information available to future researchers through the Linus and Ava Helen Pauling Papers at Oregon State University.

During several visits to Condon, Oregon, where Linus's mother's family were old settlers, we talked to several of his grade school classmates: Bob Eaton, Ernie Fatland, Walter Shannon, Anne Bower, James Burns, and his brother, Rick Parish, and to another classmate, Esther Fergueson, now living in Sunnyvale, California.

We interviewed Peter Pauling, Linus's chemist son, in London in 1964. On the same trip we also interviewed Anita Osler Pauling, his former daughter-in-law, at her home in suburban Geneva.

At Oregon State College of Engineering and Agriculture (now Oregon State University), at Corvallis, we talked at length with his professor, Dr. Samuel Graf. At California Institute of Technology, we interviewed the already-ailing Robert Corey, now deceased, and Carl Anderson.

In more recent years, we have interviewed Professors Max Delbruck at California Institute of Technology, Henry Taube and Michael Boudart at Stanford University, Jerry Donohue at the University of Pennsylvania, and James Watson at Cold Spring Harbor Laboratories. We have had numerous interviews with Arthur Robinson.

We are greatly indebted to the psychologists who volunteered their time to interpret Pauling's Rorschach Protocol: Michael Wogan, Clifford DeCato, James Kleiger, Paul Lerner, John Exner, Eric Zillmer, Vincent Nunno, and Richard Kramer.

Clifford Mead, Shirley Golden, and Ramesh Krishnamurthy of the Linus and Ava Helen Pauling Papers collection at Oregon State University were most gracious and helpful in guiding us through the mounds of documents on file at Kerr Library in Corvallis. Pauling donated his and Ava Helen's papers to Oregon State, which is the major center and resource for Pauling studies. We will contribute all original materials we have to their collection.

We are especially indebted to Luke Burke, professor of chemistry at Rutgers-Camden, for reviewing the discussion of chemical and other scientific topics. Any errors that may have crept in did so despite his patient efforts to explain some very difficult topics.

Others who were helpful in different ways were Paul Bruff, Phillip May, Leonard Lotter, Joseph Goodman, Ambrose Nichols, Jr., Hallock Hoffman, Sidney Katz, Linda Pauling Kamb, Lillian Goertzel, Stephen

Barrett, Anna Cohen, Paul Elovitz, Tom Jukes, Stephen Brown, William Lipscomb, Elizabeth Hart, and the members of the Psychohistory Forum in New York City.

Of course, none of these helpful people is in any way responsible for the interpretation we have made of Pauling's life. Some of them may disagree with us strongly on certain points. Pauling's life was long and varied, and most of our interviewees knew him for a limited period or were familiar with only one aspect of his life or work. We, alone, are responsible for the way in which we have put these pieces together into a coherent picture.

We are grateful to Gwen Goertzel for the drawings illustrating Pauling's scientific accomplishments, as well as for her encouragement and support.

INTRODUCTION

As my father's son I am certainly among his fondest admirers. Now I have ample evidence that all over the nation, and elsewhere in the world, innumerable people share my affection and admiration in their own ways. They not only admire but *love* him. They have read his books and articles, have seen him on television, have attended meetings where he spoke, or by chance sat next to him at a banquet or on the airplane. A number have received personal messages from him or talked with him over the phone. They feel a personal connection, a strong bond, with his values—his valiant anti-nuclear-testing stance, his dedication to world peace, his emphasis on vitamin C as a valuable nutrient for both prevention and treatment of disease, his compassion for the human condition of suffering. For a number of our supporters, Linus Pauling may be the closest embodiment, in this age of uncertainty, amorality, and constant conflict, of a living universal hero.

—Linus Pauling, Jr., 1993

If there is a need for a universal hero, it is easy to see why Linus Pauling would be nominated for the role. In many ways he seemed larger than life. While still in his twenties, he became a world famous scientist, renowned for his unprecedented success in applying quantum physics to the study of complex molecules. His theory of the chemical bond is one of the landmarks of modern chemistry. His work in chemical biology, done during the middle years of his life, was almost equally outstanding—more than anyone else, it was he who founded the new science of molecular biology. A powerful public speaker with a gift for communicating complex ideas, he traveled the world addressing admiring audiences. Pauling was also acclaimed for his activism as a citizen, although his activities were controversial. Born into a Republican family, Pauling moved to the Left during the 1930s, supporting Upton Sinclair's quasi-socialist campaign for governor of California in 1934. When the Nazi menace threatened Europe, he favored American union with Great Britain as a way of committing his country to the defense of that threatened land where he had so many friends and colleagues. Once the war began, he put his basic science on the back burner to commit himself to the war effort, designing an oxygen meter for airplanes and submarines and synthetic blood plasma for medical emergencies.

After the war, Pauling refused to be swept up in the anti-Communist and anti-Soviet tenor of the times. He used his scientific credentials to challenge the government's claims that fallout from nuclear testing was not harmful. When the Senate Internal Security Committee subpoenaed him to testify in the hope of discrediting him, he stood up to them, refusing to name those who helped him collect the signatures of scientists from around the world. He won that confrontation, forcing the committee to back down. Ultimately the government conceded the dangers of fallout and signed an atmospheric test ban with the Soviets.

But he paid a price scientifically. In 1953, he rushed to the press with an erroneous model of the structure of the deoxyribonucleic acid (DNA) molecule just before James Watson and Francis Crick published their famous and correct double-helix model. Many believe he might not have made his mistake had he seen the X-ray crystallography work of Rosalind Franklin that had helped Crick and Watson in their research. But because of his political activities, the U.S. State Department denied him a passport, thereby preventing him from going to Great Britain and seeing this important work.

The right wing of the 1950s denounced Pauling as a Communist, or at the least a fellow traveler. Pauling and his wife were, in fact, openly active in several organizations dominated by Soviet sympathizers, as well as dozens of others. But they were never apologists for the Soviet party line, or any other organizational dogma, nor did Pauling refrain from challenging the Soviets when he thought them wrong. He opposed Soviet bomb tests as forthrightly as he had U.S. tests. And when the leading Soviet chemists denounced his theories for contradicting the sacred Leninist doctrine of dialectical materialism, Pauling stood his ground, blasting them for corrupting science with political dogma. Years later, the Soviets recognized the error of their ways and granted him the Lenin Prize.

These victories over Stalinism and McCarthyism were richly satisfying for Pauling, not in the least because each was recognized by a Nobel Prize, the first for chemistry, the second for peace. By middle age, his life had been a remarkable string of successes, in both the personal and the public realms. His marriage to his college sweetheart had been happy. The couple remained devoted to each other throughout their lives, sharing the same political enthusiasms and raising four children. While their family life was not immune from conflict or stress, all of the children pursued professional careers and gave the Paulings twelve grandchildren. They had a lovely custom-built home on Big Sur, its walls meeting at angles mimicking those of the complex molecules that Pauling had first untangled.

Then, at the peak of his prestige and renown, instead of settling into a comfortable elder statesman role, Pauling shocked his friends and admirers by launching a crusade to prove that vitamin C would cure the common cold. When the evidence was unconvincing to most specialists, he went on the offensive. He claimed that vitamin C was a panacea, useful as a preventative and treatment for all kinds of diseases, including cancer. His advocacy of "orthomolecular medicine," the use of large doses of vitamins and other natural substances to regulate the body's molecular structure, made him a hero to health food enthusiasts, but it raised grave doubts among many of his scientific admirers.

In 1974, the American Psychiatric Association published a devastating critique of his claims that massive doses of vitamins were an effective treatment for schizophrenia and other mental disorders. In 1985, doctors at the Mayo Clinic published the results of rigorously controlled double-blind clinical trials that did not confirm Pauling's claims that vitamin C was effective in the treatment of cancer patients.

Writing in *Nature* in 1990, R.J.P. Williams saw Pauling's transition "from being a public figure of high stature with an idealistic philosophy, to being viewed as a lonely crank," as "a fall as great as in any classical tragedy." While Pauling quickly recognized his error on DNA, accepting it as the kind of mistake any scientist can make and more or less discarding his previous ideas about vitamin treatment of schizophrenia, he never abandoned his claims about vitamin C and the treatment of cancer, and in the last years of his life he felt vindicated when a national medical conference gave him a standing ovation, claiming that "Pauling was right all along." The megavitamin issue remains open to scientific debate, and Pauling can be credited with advocating an alternative point of view and stimulating new directions of research.

But one event remains a black mark on an otherwise rich and honorable life. The action that most saddened Pauling's friends and admirers in the scientific community was not his strident orthomolecular crusade, but his apparent betrayal of a young scientist who had been his closest collaborator and disciple. Arthur Robinson had left a promising university position to become cofounder with Pauling of the Linus Pauling Institute of Science and Medicine in Palo Alto, California. Robinson was president and research professor at the institute when he released the results of research with hairless mice that suggested that megadoses of vitamin C, equivalent to those that Pauling had been recommending to human cancer patients, might actually stimulate the growth of cancer. Pauling thought it was premature to release the results, which required further replication and study. For refusing to follow Pauling's advice, and perhaps for other reasons, Robinson was not only removed from the presidency but also dismissed from his tenured professorship. After a multimillion-dollar legal struggle, the institute paid Robinson half a million dollars of contributors' funds for libel and slander. Ironically, several years later, Pauling completed the research with the hairless mice and was able to show that the anomalous finding was apparently a statistical fluctuation just as he had suspected.

Some people who have known Pauling well for many years believed that this conflict brought to the surface aspects of Pauling's personality that had been invisible to his admiring public. One prominent scientist, Martin Kamen, told us, "My unhappiness with Pauling as a person is that he is an egomaniac, although a brilliantly intuitive scientist, and his behavior in the Robinson business was completely inexcusable."

It may be inevitable that a "universal living hero" will fail to live up to his billing if one examines his life in detail. Pauling's human weaknesses are surprising only because our expectations of him are so high. Pauling was not superhuman, but he was a very great man, with an extraordinarily sharp and creative mind, a firm inner determination, and a strong moral sense. His life holds many lessons and poses many intriguing questions for all of us.

1

End of Childhood, 1901–1913

Linus Pauling was born on February 28, 1901, in Portland, Oregon, in surroundings remote from the celebrated world of science and politics where he starred as an adult. His adult success resulted from his ability to distance himself from his childhood environment and to rise above it. His parents' preoccupation with each other and with their own problems gave him the freedom to develop his own way of life.

Linus's parents met in the isolated frontier town of Condon, Oregon, high on the plateau east of the Cascades, where most of the men worked as wheat harvesters, railroad hands, or wild horse trainers. Lucy Isabelle Darling, called "Belle," was eighteen when she met Herman Henry William Pauling, twenty-three, at a dinner party at her sister Goldie's house. Eligible men were scarce in Condon, and the young women of the town were envious when, six months after their meeting, Lucy married the wholesale drug salesman from Portland. Most of the town attended the reception and dinner at the Masonic Hall that followed the wedding. Nine months and one day later, Linus was born in Portland.

Herman Pauling was descended from Prussian farmers. His grandparents, Christoph and Catharine, immigrated to a German settlement in Concordia, Missouri, where Herman's father, Carl, was born. Carl took his family to California and then to Oswego, Oregon, where he became an ironmonger in a foundry. Herman was apprenticed to a druggist after finishing grammar school.

The story of Herman and Belle's romance was a popular topic in Condon for many years. Herman and Isabelle were very close, sharing an intense relationship to which the children were clearly secondary. In this sense, it closely paralleled Linus's own marriage to Ava Helen Miller. But Herman's career did not match Linus's striking achievements; he achieved only modest success, first as a salesman, later as an independent drugstore owner. He moved several times from community to community in search of better opportunities. These changes of scene provided Linus with stimulation and variety in his environment but also prevented him from forming the stable links to a community that were more typical of children of his generation.

Linus was a sturdy but homely baby, with the clear skin and luminous blue eyes of his father. His father's curls were buttercup yellow; the baby's were reddish gold. Isabelle had been plump in the early months of her pregnancy, and the family worried that she might be the target of finger-counting gossips in Condon if the baby's birth was premature. When her sister Goldie breezed into the Pauling apartment to help after he was born, they rejoiced together over the baby's proper timing of his arrival: nine months, and an extra day for good measure.

When Linus was three months old, his mother wanted to take him on the two-day journey by train and boat from Portland to Condon to show him off to her sisters, father, and friends. Herman agreed, although he feared the journey would be too tiring for the infant and nursing mother. It was. They arrived in Condon exhausted and ill. Belle did not find the strength to write Herman of her safe arrival as he had asked her to do. Herman, frantic with anxiety, made an extravagant long-distance telephone call, then sat down to write a letter on April 25, 1901. The letter conveys much of his feeling for his wife and child.

My dear wife and baby,

I feel much better having heard from you. Sorry you and Linus have not been well. I was awfully worried about you. I would have phoned before, but for Mrs. Smith who kept saying that you were all right. Of course, I knew that if it were serious you would have phoned before. Take good care of yourself and our dear little boy. Kiss him for me. I will send you anything you want if you will only write. Well, dearie, do write and tell me about your trip. Have Dr. G. [Dr. Gillette] see the baby often. . . . Don't take any chances for the little fellow is too precious to lose. I am well, but lonesome. It is the same old story. We don't miss you until you are gone. . . .

Lovingly,
Your Hermie

Portland was more diverse and stimulating than Condon, and Belle and Herman Pauling were small fish in a big pond, living in a one-room apartment in the center of town. On their walks with the baby through various neighborhoods of the city, they passed newly built mansions, each occupying a full city block. Linus learned to recognize as landmarks the iron deer on the lawns and the black "iron boys" who never moved from their posts by the watering troughs. They lived near the fountain presented to the city by Stephen Skidmore in 1888, and the baby liked to watch the water that poured from the basin held aloft by bronze maidens. They also lived near the exotic Chinese neighborhood, which had a long wall plastered with bright strips of orange paper that had messages and notices written in bold, black characters and flapped in the wind. The Chinese drugstores sold pairs of dried turtles meant to be boiled and eaten as a cure for rheumatism. The beads of the abacuses clicked as customers paid for strange and salty foods drying on round straw mats: shark fins, little devilfish, oysters, shrimp, mussels.

Herman and Belle encouraged Linus to explore this environment. Herman made friends with a Chinese merchant, who taught Linus to count to one hundred in Chinese when the boy was only two. The men took great delight in the boy's quickness and praised him effusively. These childhood experiences may reflect Linus's lifelong facility for learning languages: he became fluent in German, French, and Italian. Linus was exposed to German as a small child when he visited his father's relatives in Oswego, seven miles from Portland, where German was the language of his indulgent grandmother.

Linus did well in Portland, eagerly sharing his parents' enthusiasm for its sights, sounds, smells, and tastes. His mother was delighted with the city after her small-town upbringing. The roses were redder in Portland, the hollyhocks taller. There were many urban attractions, combined with a small-town friendliness. Cows were often tethered in the back yards where the housemaids beat the velvety carpets on clotheslines.

Freedom meant some dangers for a small child. At twelve months, Linus was badly scalded when he spilled hot coffee from a pot on the stove near his high chair. This was his last serious accident. He learned to be wary and observant but didn't lose his curiosity. At three he was an independent, self-sufficient little boy, older than his years.

When Linus's sister, Pauline, was born, however, Herman and Belle began to think seriously about moving out of the city. They were too crowded in the small apartment, and Herman could not afford better housing in Portland. Belle stayed with her husband's parents, Charles and Adelheit Pauling, in Oswego, while Herman looked for new housing for the family. He decided to travel again for the Skidmore Drug Company and to make Salem, Oregon, his family headquarters. This plan was satisfactory financially. Herman sometimes sold as much as a hundred dollars worth of drugs and notions to a single retailer in Forest Grove, Sheridan, Beaverton, Dayton, or McMinnville. But Belle was not happy alone with the children, and he missed her.

Finally, Herman gave up his salesman's job and set up his own drugstore in a new bank building in Oswego, his own hometown, where he had attended grammar school and been an apprentice to the druggist. The family would be happy in Oswego, close to his parents, but the business climate was poor. Oswego had suffered badly from the depression of 1894, and the ironworks that was to make Oswego the Pittsburgh of the West had closed down. Many jobless workers were forced to abandon their little company houses and leave town. The only doctor had left; the dentist was killed in an accident, and no one took his place. The druggist, too, had moved, and Herman was the only health practitioner left to serve the residents.

Herman's parents were delighted to have their only son and his family in Oswego, and they all enjoyed the beauty of the Willamette valley and the joys of quiet family life. Grandmother Addie Pauling liked company. Her coffeepot went on the fire when the front gate clicked its announcement of company coming. She fed her grandchildren sweet buns and butter cake while their grandfather held them on his knee and told them stories of his boyhood back in Missouri. Linus was quick to learn German jingles from his grandfather and grew accustomed to hearing the cats, chickens, and cow admonished in German.

The Paulings rented a neat little frame house with a picket fence. Linus's second sister, Lucile, was born in Oswego and Linus became immersed in the world of his mother and sisters because his father was always busy with his business. Herman Pauling was the only member of the working-class Blanken-Pauling family to start a business. He threw himself into it and was a great success with the local residents, who were in desperate need of medical services. Word spread that Herman Pauling was as good as any doctor, and he charged no fee for a consultation. He often visited bedridden customers to see how they were doing and give them a little attention. All his life Herman had a genuine concern for public health and a curiosity about the effects of his drugs on his customers.

None of this produced financial success, however, and Oswego was not really able to support a druggist properly. Word came from Condon that the long period of drought was over and the wheat fields had turned to gold. The railroad had been extended to Condon, and business was good. Herman decided to go to Condon to arrange housing for the family and to set up another drugstore there.

The move to Condon put young Linus in a small town where his mother's family were among the most respected members of the community. Linus Wilson Darling, his grandfather, had served as teacher, farmer, surveyor, postmaster, and lawyer; two uncles were proprietors of the general store. The move across the Cascade range from the lush Willamette Valley was to a small market town dwarfed by the grandeur of the high desert of eastern Oregon. Condon was built on a high plateau; on a clear day a boy could see the Blue Mountains to the east and the Cascades to the west. Canyon walls rose two thousand feet above the John Day River along the road that wound its way to the town of Fossil, a few miles south of Condon. The high desert country was arid, wild, and lonesome. There were no trees in Condon and the houses were unpretentious wooden two-story structures set in straggling rows along wide dusty streets that became mudholes during the winter rains.

Towns like Condon were settled by people intent on beginning new lives. The Darlings, Linus Pauling's mother's family, were no exception.

Orphaned at eleven, Linus Wilson Darling, Linus Pauling's grandfather, apprenticed himself to a baker who made him sleep in a barrel and refused him time to go to school as promised. His mother had told him on her deathbed that he should learn as much as possible since no one could take away what he had in his head, so he ran away from the baker and headed west. By 1871 he was settled in rural Oregon and had somehow learned enough to teach in a country school. His older brother William, Jr., joined him there and the rest of the five orphaned children kept in touch and retained a strong sense of family identity. Linus Wilson Darling met and fell in love with a pretty young woman from Turner, Oregon, near Salem. He followed her to her hometown, taught the rural school where she was enrolled, and married his favorite student when the school closed in the early spring.

Darling pursued many careers in the context of small frontier towns, where formal professional qualifications were not always necessary. He was a lackadaisical farmer, but an enthusiastic surveyor and a competent postmaster. He studied the law in his spare time and became a licensed member of the bar. He was called "Judge" by his neighbors for decades.

From 1905 to 1909, Linus Pauling lived in Condon in fairly close proximity with his maternal grandfather, whose custom it was to listen to his daughters read aloud from books such as *Pilgrim's Progress.* For years, Linus Pauling's mother and his three aunts were in awe of a great book called *Blackstone* that their father read and reread. Not one of them dared even to touch it. Each of the girls was named after a queen, and each was frequently admonished to act like one. Since their mother had died young, their father was the dominant influence on their lives.

The similarities between Linus Pauling and the grandfather whose name he bears are striking. Linus Wilson Darling had a strong scientific curiosity. Whenever he planted a crop, he kept detailed records of the seeds used and the results obtained. He sent to eastern nurseries for seeds of the box elder, locust, ailanthus, 'Royal Ann' cherry, sugar maple, and Scotch pine to find out whether any trees could be successful in Condon, in high desert country where rainfall was unreliable. He planted several varieties of vegetables and on walking tours throughout Gilliam County informed neighbors of the results of his experiments. He kept detailed drawings of his construction projects and was in great demand as a teacher and public speaker. He also wrote letters and informative feature stories for the local newspapers.

"Judge" Darling's diary captured much of the flavor of frontier life, as well as of his own personality. One Sunday afternoon, August 18, 1888, he recorded the following events:

> Along in the afternoon, a row took place in town [Condon]. A man named Barry, a Swede, got gloriously full at Watson and Barr; a saloon across the street from my store, and then he tried to quarrel with everyone. Drawing his revolver on Jessie Mulcare, that person took it away from him, then I went over to where the quarrel took place. Mulcare handed me Barry's revolver and went down on him and none too quick for he lunged at me and cut through my coat and vest. I put him under $400 bond.

When the man appeared before him for trial, Darling fined him lightly and let him go. He usually had to go on living in close proximity with the people whom he judged, a factor that could involve some risk. In one case he noted:

> Hall vs. Ward came up for trial today. Neither party would call for a jury and I shall have to give a decision between two of my neighbors. To decide against Ward means to incur the displeasure of about five families of "bulldozers" who have no regard for law, order or decency.

He decided against Ward.

The diary also recorded personal matters, with less detail and attention to emotions. Perhaps Linus Wilson Darling found it less painful to examine his feelings less closely than he did his scientific and judicial pursuits. The two great tragedies of his adult life were recorded as follows:

> Friday, July 17, 1888. My first-born son came into the world silently and was taken to the Condon cemetery for burial.
>
> Saturday, August 11, 1888. A sad day in my house, for the light went out of the eyes of Alice Delilah Darling (wife) forever.

As an adult, Linus Pauling most clearly resembled his mother's family when he spoke to an audience. Intent on capturing the attention of his listeners, he was a compelling figure with an animated face and a persuasive smile. He was never as flamboyant as his Condon relations, however, and off stage he was more like his more stolid Pauling relatives. In the laboratory setting, he was composed and disciplined, reluctant to engage in idle conversation, and somewhat cool to anyone who might want to interrupt his work.

Linus was an active four-year-old when he moved to Condon in 1905. His grandfather had passed the bar examinations and was an attorney-at-law. The Condon *Globe* carried his advertisement, a simple two-inch announcement of his name and profession, nothing more. Linus's father had still to establish himself, however, and launched his new drugstore with an advertising campaign, unprecedented in Condon. The first advertisement announced the opening of the Pauling Cash Cut-Rate Drug Store. The special for the week was a $1.00 family medicine chest worth $2.50, which included a box each of salve, liver pills, cold cure, headache tablets, worm lozenges, tooth powder; a cake of glycerin soap; and a bottle each of white pine and tar cough syrup. Prescriptions could be filled for 25 cents apiece.

Each weekly issue of the *Globe* carried a new full-column Pauling advertisement. One ad was a six-hundred-word description of a blood purifier, six bottles for five dollars. He announced a ten-day sale at the store where "Pauling's Prices Please the People" and continued in the alliterative vein with "Pauling's Pink Pills for Pale People."

In the March 30, 1906, edition of the *Globe* he offered twenty-five-cent bottles of "Dr. Pfunder's Oregon Blood Purifier." This seemingly miraculous compound of natural herbs, roots, barks, and berries was not only "an almost certain cure for all diseases arising from derangements of the liver and kidneys, or an impure condition of the blood, such as scrofulous affections, erysipelas, pimples, blotches, boils, salt rheum, ulcers, cold sores, dropsy and dyspepsia," but a rheumatism, biliousness, coated tongue, and bad breath remedy.

Dr. Pfunder's blood purifier, containing only natural ingredients and allegedly carefully analyzed to assure the proper dosage, could be viewed as a primitive example of "orthomolecular medicine" as Linus Pauling later defined it. It sought to treat disease and maintain health by regulating the concentrations of substances that are normally present in the body. Although vitamin C was not isolated chemically until 1928–33, the berry and grape juices Dr. Pfunder used may have included significant quantities of it.

Linus was an alert, observant youngster who was clearly impressed by his father's and grandfather's activities. He was also absorbed in the newly booming life of the small frontier town. Elevated wooden platforms made a sort of sidewalk in front of the Main Street stores, and Linus and his cousin, Mervyn Stephenson, crawled under them to recover coins dropped by careless shoppers. At five years of age, Linus wrote a love letter to a first-grade girl that was published in the *Globe.* His picture in chaps and a sombrero appeared on a postcard that could be purchased in the stores. The caption was "A Condon Cowboy."

This appellation was not totally inappropriate, since Linus spent a good deal of time at the Stephensons' ranch twenty miles out of town, where the hands found it a great sport to introduce him to country ways and let him shoot their hunting rifles. They also had a pet mare, Old Jenny, who was patient when all the children piled on her at once with Mervyn, the oldest boy, at the head of the row, followed by Lilah and Pauline, Rowena, Lucile and Babe, and Linus at the rear to help hold the little girls.

Most of the time, Mervyn and Linus avoided playing with their younger sisters. When in town, they kept getting into mischief, taking crackers and sardines from the Stephensons' general store to eat in the alley. They once cut corners from plugs of tobacco and chewed them and were spanked, not so much for chewing tobacco as for ruining the plugs, which were no longer fit for sale. They were spanked again for sucking on rubber siphons that hung out of the kegs of port wine sold "for medicinal purposes only." Siphons had replaced corks or wooden bungs because they permitted less leakage. To start the flow it was necessary to suck on the rubber tube, a temptation that Linus and Mervyn were unable to resist until they became ill from overindulgence.

On another occasion, Linus accidentally tumbled into the cesspit of the privy in back of the saloon. Mervyn was unable to reach his outstretched arms. Luckily he turned a complete somersault while falling and landed feet down, ankle deep in the stinking mess. His father came running to pull him out and poured bucket after bucket of water over him.

Linus had a happy and ordinary childhood in Condon. School was neither challenging nor especially unpleasant. He attended a one-room school for first grade, and then the larger grammar school where his parents had courted on the front steps. There were two grades to a room, and most of the teachers were kindly, energetic young women who had little formal education. There were few manufactured toys, and the boys saved string to make baseballs and played "Duck on the Rocks." Both boys and girls played hide-and-seek and dare-base at recess or during the long noon hour. For Christmas, Mervyn and Linus got new train sets, while the girls found trinkets and new dresses for their dolls in their stockings. They all strung popcorn and cranberries to decorate the family Christmas trees.

Linus had only one expensive purchased toy, a wagon that on one occasion he and some other boys used as a base to make a toy threshing machine. An empty paper coffee barrel placed on its side in the wagon bed simulated an engine. A smaller can fitted into the coffee barrel was the smokestack. When the experiment ended, it was necessary to hide the wagon in the bushes since the fire the boys had lit in the can had scorched the bottom of the wagon. When the shock of what they had done wore off, they pulled the wagon out cautiously. No adults noticed the damage.

In the small town environment, children were not separated from adult life. Linus and Mervyn mingled with the harvest hands, taking them water and listening to their stories. When a boy ate well and was not ill, his elders were usually too busy to pay him much mind. When company arrived from a distance, it was a major event that brought the families together. Uncle Herman Blanken visited from Oswego and was fascinated by the thousands of acres of undulating wheat fields. Herman Pauling dearly loved the uncle for whom he was named, though he noted that "he was his own worst enemy," and worried that he had inherited his uncle's poor business judgment.

Children in Condon created their own activities, and Linus was much like the other boys in town. His cousin Mervyn's comment is typical of his contemporaries' memories of Pauling:

> I would say that he was more normal and average than most boys. I never felt anything special about him, but I was not the age to think about things like that. We played out until dark, but he still found time to read widely and avidly. He was not compulsive about being good, and he was not a mother's boy. He was not awkward, but he was not an athlete.

However, Linus did make an effort to stand out and to draw attention to himself. On one occasion, he joined a group of children who slipped into the Union Dance Hall to play. There was a stage at one end of the hall, and they decided to put on a show. First the boys watched while the girls took turns singing, reciting, or dancing, then the girls watched. Linus danced out on the stage like a clown in the silent movies. As he left the stage, he let his pants drop to his ankles, pretending that he was terribly embarrassed by the failure of his suspenders. The children laughed, clapped, and stamped so loudly that one girl's father heard them from the store below and promptly threw them out of the hall.

Linus was only nine when the carefree years of childhood came to a sudden end. His world was shattered by a series of incidents beyond his control. Condon went into a doldrums, following an initial boom caused by good weather and the coming of the railroad. Portland, on the other hand, was booming, thanks in part to the impetus given by the Lewis and Clark Exposition in 1905. Linus Wilson Darling then upset the whole community by divorcing his well-to-do second wife and marrying a young schoolteacher of his daughter's generation. He moved with her to Portland, where he started his law career anew.

Belle Pauling grew restless in Condon and longed for the Portland of her honeymoon days. Condon was too small to absorb Herman's talents for merchandising, and he decided to build a drugstore of his own in the more competitive atmosphere of the larger city. The move to Portland was a lonely and threatening experience for Linus. He turned more and more to his books for solace after losing his Condon playmates. He had already read and reread *Ivanhoe, Alice in Wonderland,* and *Through the Looking Glass.* "What shall I read now?" he asked his father, who was not too busy to be aware of his son's exceptional intellectual abilities.

Herman was not a bookish man himself and handled the matter by a letter to the editor of the Portland *Oregonian.* He was the father of a nine-year-old boy who read far beyond his years. What books would be appropriate for such a boy's library? Herman's letter reveals an eagerness to share his pride in his son with the community at large, rather than directing his question to a librarian or teacher.

Belle had more leisure now that her children were in school and enrolled in a German class in the high school. Although she loved her kindly mother-in-law, it irked her not to understand what the German-speaking women of the Blanken-Pauling family in Oswego were saying when she visited them. Her father, still a curious and energetic scholar, also began the study of German. Clients were few and far between at his new office, and he needed something to make him look impressively busy when one did drop in. He also wanted to offer Belle some competition and encouragement.

Just when it appeared that the relocation to Portland was beginning to work out, Linus Wilson Darling died of a heart attack at the age of fifty-five. His condition was complicated by nephritis, which Linus Pauling also experienced in middle age. The struggling Portland lawyer left a childless young wife, four daughters, and seven grandchildren. The daughters had been hurt and annoyed when the young wife tried to influence her husband to make her, rather than his daughters, the beneficiary of his life insurance policy. He did not do so, and now they were guilt-stricken as well as bereft. Belle was depressed by her father's death, and her melancholy made Herman even more tense and anxious as he went about trying to establish his new business.

There was no reason, Linus Wilson Darling's daughters decided, for his untimely death to deprive the grandchildren of a holiday. The Portland Rose Festival was on June 10 that year. Goldie Stephenson traveled from Condon with her four children. Aunt Abbie and Aunt Stella, single and working, helped the young married sisters take the seven lively children, aged four to eleven, to the festival. Linus was nine.

Herman chose to stay at the store, but he was pleased to see Belle and the children enjoy themselves. He had not been well for several months and had recurrent sharp abdominal pains. Since they tended to disappear when he ate, he neglected seeing a doctor. Feeling restless and in pain when he went home to his empty house, he ate heavily from a pork roast that Belle had made for a sandwich supper when the family returned from the festival. Herman felt better after eating and returned to the store but collapsed there in agony and was carried home.

When the tired merrymakers returned at dusk, Abbie Darling, a trained nurse, saw at once that her brother-in-law was dying. The family doctor was at the festival, and a frightened young physician was called. He gave Herman a sedative that masked the pain but failed to order the emergency operation that was needed. Herman Pauling, aged thirty-three, died of a perforated gastric ulcer on June 11, 1911.

Linus learned of his father's death by hearing a neighbor say, "These poor little children do not know what it means to have a father who is dead." He wanted to correct her, to tell her that he had long since understood fully the meaning of the word *dead.* He accepted the brutal reality of his father's death as a sad but natural fact. People were born, grew up, worked, and, sooner or later, died. Although he and his sisters had always known their parents loved them, they also never doubted that they loved each other best of all. Their loss was secondary to hers.

In middle age, Pauling recalled, "You know when my father died my mother became hysterical. As we went on the streetcar down Hawthorne Avenue to take the train to Oswego where papa would be buried, where the funeral took place, I remember that I was upset by my mother making a disturbance. This was an attack of hysteria that I suppose is not uncommon."[1] Belle's reaction may have been quite normal for a young mother whose husband was about to be buried, but it was understandably frightening to a nine-year-old boy who had suddenly become the man in the family.

Linus longed for an end to the coming and going of the head-wagging, hand shaking strangers who visited the family to mourn Herman's passing. He had little understanding of the new burdens his mother had to assume as head of the family either then or later, or of the effect the death would have on his own life. Until Herman died, Belle had never paid a bill or discussed money matters with her husband. In her childhood home her father and eldest sister, Goldie, had made the important decisions. She used the insurance money from her father and husband to purchase a rooming house on Hawthorne Street in Portland. She desperately wanted a safe refuge where she could live out the rest of her days and keep a roof over her children's heads.

When the bereft Paulings had lived only a few months in the new house, Belle was ordered to bed by her physician, with a diagnosis of pernicious anemia. After several weeks she was able to dismiss her hired housekeeper, but she had, henceforth, the dreary conviction that she might not live to see her children married in homes of their own. Her healthy, long-limbed children didn't wonder too much why their mother was so often tired, pale, and breathless. Women, they knew, often had mysterious ailments that they discussed among themselves and from which they always recovered. Belle had no visible wounds and protected them from her own fears and anxieties.

Pauline and Lucile had plenty of playmates in the new neighborhood. They played long games of run-sheep-run in the dusk on summer evenings, huddling and giggling in the shadows until the mysteriously worded signal of their chosen leader would send them running pellmell to the goal. Linus, however, held back and seldom played outdoors, apparently preferring solitary intellectual activities. He said in an interview in 1964, "As I think of myself as a boy I conclude that I was more interested in being by myself than with being with other people." This preference persisted in his adult life, as he observed, "Ava Helen knows that I used to complain, even still do, whenever we have an engagement that requires that we be with other people."

On rainy days, of which there were many in Portland, Linus often joined his sisters at whist or bingo or lotto with their friends at the kitchen table. But he made few friends of his own. Belle was adamant about their being clean, well spoken, and honest but she did not attempt to supervise their academic or social lives. They were free to make their own mistakes, to seek part-time jobs on their own, and to choose their pastimes or companions. Linus often preferred to be alone.

Household storms were infrequent, if only because everyone went his or her own way. But Linus was determined to assert his independence from his mother and sisters. Once Lucile stole a ride on the bicycle of a boy she knew and went home with bloody knees and copious tears when she was pushed to the ground. Belle commanded Linus to go immediately to beat up the boy who had hurt his sister. Linus was unmoved, logical, and indifferent. He thought that the boy involved was a harmless, decent fellow and wasn't sure Lucile merited a defense. Pauline stoutly championed Linus while Lucile, who was attracted to the boy whose bike she had borrowed, was not so eager as her mother for revenge. Belle, however, was bitterly disappointed by Linus's indifference to his sister's injury.

All three living participants in this argument, interviewed separately, agreed that it was a memorable event. The sisters wonder whether Linus was developing the independence that served him so well as an adult when he made decisions according to his own sense of values, disregarding the opinions of others. Or was it related to his hatred of war and violence as an adult? They agreed that their usually mild and patient mother was truly angry with Linus when he would not defend his sister. Perhaps she was frustrated at his inability or unwillingness to take over from his father and grandfather as the male head of the household.

Belle Pauling tried her best to make a good life for herself and her children. She fulfilled one of her own childhood dreams when she saved a hundred dollars and bought a fine Kohler piano. Belle was often too busy or ill to practice and Pauline didn't take to the piano, but Lucile was soon good enough to have pupils of her own. Linus was always seeking part-time jobs. On his bicycle he delivered special delivery letters for the post office. For two two-week excursions to the seashore during consecutive summers, he earned money by setting pins at the bowling alley. All the children dug clams, which Belle made into clam fritters.

Linus made weekly trips to Oswego to see the Pauling grandparents, who were beginning to seem old and frail to him. Sometimes his mother and sisters accompanied him; often the children took a picnic basket and hiked to the lake a mile away. Linus was also in touch with other relatives. His mother sent him to visit his paternal Aunt Anna and her husband, Judge James Ulysses Campbell, later chief justice of the Oregon Supreme Court, in Oregon City. The judge's home provided more intellectual stimulation, including a new set of encyclopedias that Linus devoured eagerly, piling them up around him like building blocks. He continued to love encyclopedias through his life. Belle did not get along with her sister-in-law, Anna, who blamed Belle for Herman's ulcers and, thus, for his death. Anna thought Belle's desire for luxuries had pressured Herman to achieve financial success.

Linus maintained a studious indifference to these delicate nuances of family relationships. The boardinghouse in Portland drew frequent family visitors, but Linus remained aloof from the series of romances, weddings, divorces, births, and even deaths around him. He maintained a similarly aloof attitude at high school, where he had reached the ninth grade by the age of twelve. He was not popular or athletic and was never elected to a class office. His teachers were not unkind but did not single him out for special attention despite his apparent giftedness. He was a nondescript boy who was always a scholarly onlooker at school, quietly making plans for success based on his intellectual abilities. He sought no particular help or guidance from teachers, relatives, or other adults.

He had one close friend, Lloyd Jeffress, a quick-spoken, intense lad who was also an orphan. Jeffress was the only person with whom he could really discuss personal matters, and they talked endlessly about sex, chemistry, Greek, jobs, and career plans. Lloyd doubted that Linus would be an engineer, insisting instead that he would become a professor. Lloyd also planned to become a scientist and as an adult became a psychologist who did research on hearing.

Lloyd lived with an aunt and uncle, who accepted Linus as readily as they did their nephew and welcomed him into a household where there was much talk about books, politics, ideas, music, and art. Linus had never experienced this kind of home environment, with an easy give and take of ideas and a love of learning. At his own home, the men who were his mother's roomers talked of business or other practical affairs. They were not indifferent to Linus but saw him as a bright, hard-working boy who needed to work as soon as possible to help his mother and orphaned sisters. Belle, also, was eager for whatever help she could get from Linus or his sisters and had little sympathy for his dreams for the distant future.

Linus and Lloyd were enthusiastic collectors of insects and minerals, although the Willamette Valley was a poor place to find interesting minerals. About all they could find were agates, and perhaps some zeolites. They got books on insects and minerals from the library and absorbed information that went well beyond what they could observe on their own. Linus copied tables of the hardness and other characteristics of minerals from books and pasted them up on the wall of a workroom that he built in a corner of the basement of his home.

He also received science instruction in high school, beginning in the ninth grade. Fifty years later Pauling remembered his teacher, Pauline Geballe, vividly.

> When I was twelve I had a course in high school, my first year, in what was called "physiography," which was general science. This interested me very much. I can still remember my teacher, an old woman, about twenty-four, I suppose. . . . I can remember her telling us what flint was, and jasper, and this of course encouraged my interest in minerals. I remember her carrying out an experiment to show the pressure of the atmosphere. She had a Log Cabin Syrup can with a little water in it. She boiled the water for a while, screwed the top on, and then let it cool until it collapsed.[2]

Chemistry was not covered in the physiography course, however. He was introduced to the subject by his friend Lloyd Jeffress, who demonstrated some experiments with chemicals that he had in his bedroom. Jeffress mixed potassium chlorate and sugar, then added a drop of sulfuric acid to start a reaction that liberated the water and produced carbon. Pauling recalled that three of the experiments "really astonished me; they pleased me immensely."[3] Enthralled, he decided to become a chemist. His grandfather was the night watchman at the Oregon Iron and Steel Company, which had been closed for fifteen years. A building whose roof had fallen in had a small chemical laboratory, from which Linus periodically "borrowed" chemicals.

His mother and sisters took little note of his scientific hobby, except when he filled the house with noxious odors or when the authorities arrived to complain that he and his friends had caused a loud noise by placing something on the streetcar tracks that exploded when the train ran over it. Belle tried to involve Linus in making family decisions, but he evaded her. She often had to speak to him four or five times to get his attention.

This resistance to his mother was part of a general pattern of avoiding close relationships with adults. Linus was more comfortable with boys his own age, because they treated him as an equal and did not try to tell him how he should run his life. Even so, he sought out only those who shared his interests. Of twenty teachers and classmates in his high school when he was there, only three remembered Pauling. Even they thought of him only as a quiet boy who was a good student. Pauline Geballe, the teacher who had introduced him to science, could remember many of the more troublesome students of his generation but had no recollection of her most famous pupil despite the fact that she was quite familiar with his adult accomplishments. Linus was not mentioned in her diary for that year, nor was he one of the students who prepared science notebooks for the school fair. He did not join the high school science club or any of the other clubs.

Linus was gifted with a quick mind, tempered with the work and personal habits of his family. At age nine he was a well-adjusted, outgoing young lad with good social skills and a flair for attracting attention. He had a secure place in a small, tightly knit community and could look forward to a successful future in whatever profession he chose to pursue.

The tragic events of his ninth year had a devastating impact on Linus's psychological development. He never really allowed himself to express the pain he felt after his grandfather's and father's deaths, perhaps because his relationship with his mother was not close enough to give him a feeling of security. Her own depression and ill health, coupled with the unfamiliar practical problems of supporting the family, made it difficult for Belle to be attentive to her son's emotional needs. She was never as close to him as she was to her daughters.

From nine onward, Linus channeled his energies into his hobbies and into part-time jobs taken to contribute to the family's support but also to give him a degree of independence from his mother. He was fascinated by the natural sciences and discovered that he had a natural aptitude for academic work.

The preoccupation with science may have had its origins at least in part in a need to sublimate emotional distress, but he was also good at it and realistic enough to recognize that scientific achievement could be an avenue to professional security as well as an absorbing escape from the rigors of everyday life. Whether through death, illness, or insensitivity, adults had let him down. He was determined to make his way on his own.

By the age of twelve, Linus Pauling had already developed many of the behavior and personality patterns he was to maintain throughout his life. He was introverted, intent on pursuing his own interests, and oblivious to conflicting demands from those around him. Emotionally, he was most comfortable when he could rely on a close relationship with one person for intimacy and support. The first special person was his friend Lloyd Jeffress, the second his wife, Ava Helen Miller. His marriage to Ava Helen closely paralleled that of his own parents, in its greater emphasis on the parents' relationship with each other than with their children. It was a traditional marriage, in which Ava Helen devoted her energies to her husband's career and the care of their children.

Linus's religious beliefs were also fixed at this stage of his life. His parents were both nominal Protestants, with little interest in religion. His father's mother, however, was an active Lutheran and even had a minister conduct monthly services in her home. When he was visiting in Oswego, Linus listened to some of these sermons, which raised his curiosity about religion. One day, while reading Dante's *Inferno* in bed at his grandmother's house, he thought he saw a halo around Jesus' head in a picture on the wall. He continued to watch and concluded that he could see a similar halo if he stared at any dark object in the room. He reached the conclusion that this was "a general physiological phenomenon," perhaps caused by an "after-image or fatigue on the retina." He thought about this more and concluded, "All the other phenomena that are discussed in Sunday School and the Bible . . . were just misrepresentations of natural phenomena." He kept this conclusion to himself, however.[4]

The men who were the young Linus's models were independent, small-scale entrepreneurs who did not let traditional barriers prevent them from achieving their own goals. The parallel between his grandfather's running away at eleven to seek his fortune and Linus's rebellion from his mother's demands in order to pursue a college career is clear.

2

Dropout of the Class of '17, 1913–1922

By the middle of his high school years, Linus Pauling had firmly resolved to live his own life as he chose regardless of what his mother wanted. His father and grandfather had lived independent lives as small businessmen, and Linus thought that he could do the same while pursuing his fascination with chemistry if he opened a business as a chemist. He and a slightly older friend, Lloyd Simon, printed up business cards announcing that Palmon Laboratories, directed by L. E. Simon and L. C. Pauling, chemists, was open for business at 1114 East Taylor Street. The name of the laboratory combined their two last names. They set up a laboratory in the basement of Lloyd's home and approached local dairies offering to do butterfat sampling at very low prices. Their home built centrifuge and laboratory skills were equal to the task, but the dairymen were unwilling to entrust their business to them because of their youth. The business failed.

Linus desperately needed an independent source of income, and felt he was past the age for running errands on his bicycle. He found a job at a local movie theater, where he worked in the projection room from seven to ten on Friday, Saturday, and Sunday evenings for $1.00 a session. He soon learned to thread the machines and tend the arc and became a full-fledged projectionist, receiving $1.25 an evening. When something went wrong with the light or the machinery he was bathed in his own sweat with anxiety.

He knew that a career as a chemist would require further academic training, including a college education. He enjoyed all his physics and chemistry courses, profiting from his study of the subjects for a year and a half on his own. He formed a good relationship with the chemistry teacher, William Greene, who was so impressed with Linus that he gave him credit for an extra semester's work. Under Greene's direction, Linus earned extra credits by doing some experiments, making preparations, and studying qualitative analysis. He stayed after school to help Greene measure the heat of combustion of samples of coal and fuel oil purchased by the Portland schools.

Linus overcame his reluctance to form relationships with adults so long as the adults were supportive of his plans. He formed a friendship with an elderly gentleman, Mr. Yocum, who lived next door. Mr. Yocum was a retired mountain guide but had an interest in the classics and easily convinced Linus that no real scholar could be without a knowledge of Greek. Under Yocum's tutelage and encouragement, Linus learned what was roughly a year's course in Greek. In the textbook that Mr. Yocum gave him and Linus cherished all his life is a written promise to master the volume completely within five years' time. Linus studied Greek as he rode the twenty-five minutes on the steam train to Oswego each Sunday. Mr. Yocum also gave him discarded bits of chemical apparatus from a laboratory where he did part-time work.

School was easy for Linus, and his grades were consistently good. He chose courses that interested him and was permitted to take more than the usual number of college preparatory courses because of his high academic standing. In the fall of 1916, he was a fifteen-year-old senior. He already had enough credits to enter Oregon Agricultural College at Corvallis, although he did not have credits for American History I and American History II, which were required for a high school diploma. He simply wasn't interested in these courses and hadn't taken them. He proposed to take them concurrently during the final semester of his senior year so that he could graduate with his class in June. The principal refused even to consider Linus's plan. American History I and II were never taken concurrently, and that was that. Faced with this obstinacy, Linus simply registered for other subjects and made up his mind to leave school in June without a diploma.

Although this may seem a remarkable act of rebellion for a fifteen-year-old boy, it was not noticed by the crusty principal and other officials of the school. Linus was an exceptional student, but so were many others, and he was not an athlete, class officer, or member of a family who would raise a fuss. Many boys dropped out of high school to help support their families. Linus was not about to let the principal interfere with his plans, but neither did he make a point of confronting his authority. He knew he could enter college without the high school diploma.

The real problem was saving enough money to cover his expenses. He worked part-time at a grocery store for eight dollars a week, but his mother was eager for him to leave the odd jobs and take a better-paying position that would lead to a career. As soon as he finished high school, she arranged for him to have an interview with a Mr. Schwietzerhoff, co-owner of several manufacturing corporations in Portland. He was hired as an apprentice machinist, a job that paid forty dollars a month. Pauling was a good machinist, and his pay was raised to forty-five dollars by the end of the first week and to fifty dollars a month at the end of only one month on the job. His mother was delighted.

Life as a machinist was dull and unsatisfying. His mother awakened him each morning at 6:15. He left for work at 6:55 and began work at 7:30. He swept out the office, dressed in his greasy overalls, and worked until the half-hour lunch period, when he gulped down a sandwich and milk while reading a magazine. He then worked until five in the afternoon with compulsory overtime three nights a week, the company paying an extra four bits for dinner. The work was boring and repetitive, and Linus felt a silent panic over being pushed into a career that threatened his life goal of becoming a chemist.

His application to the Oregon Agricultural College (OAC) was not taking its normal course. The secretary at Washington High School had made out his transcript inaccurately, leaving out several of his courses. The principal was on vacation and the necessary correction was held up interminably. Each day he was disappointed when the mail brought no admission slip from OAC; his mother was happier and happier. Linus was earning a man's wages and taking them home to her and his sisters.

In a frantic effort to be free of the machine shop, Linus plunged into business in his spare time after working hours. He and two friends built a photography laboratory in one of their basements and went after the business of a local photography firm that was under new management. They hoped to get a contract as official photographer at Washington High School. At least photographers worked with chemicals, and Linus thought he might discover a way to extract something valuable from the expended chemicals. He hoped to get the business established before going away to college and then to collect income from it, as much as five dollars or ten dollars a week, while he was away.

The photography business succeeded to the extent of remaining solvent, but Linus soon saw that it would not be enough to contribute substantially to his college expenses. Nor did he really want to remain in Portland and be a photographer, although it was preferable to working in the machine shop. His mother was increasingly angry over his insistence on going to college. She was in poor health and found it difficult to operate the boardinghouse. She considered going to college effete and strange: A real man was self-taught, as her father and husband had been. Linus could study to be an engineer while he worked at the machine shop and be well paid at the same time.

These pressures were hard on Linus, who also had to deal with all the normal problems of adolescence. He felt old beyond his sixteen years and filled his diary with doubts:

> The more I look in the mirror, the more peculiar my physiognomy appears to me. I do not look at all attractive, but I am a prejudiced judge. I already have faint horizontal wrinkles in my forehead and my upper lip projects to an unnecessarily great extent. I must remember to restrain it.

Linus regretted not graduating with his classmates, particularly since he was unsure he would ever go to school again. He also worried that he wasn't prepared to handle the courses at Oregon Agricultural College if he was admitted. He was afraid that he would be too young and inexperienced to compete with his larger classmates. To build his self-confidence, he continued his studies, discovering a new way of learning Latin by memorizing Latin poems and their English translations.

Still, Linus felt that he had committed himself by dropping out of high school without the diploma. He never really wavered in his determination to go to college and sought out support from adults who agreed with him. He was particularly encouraged by Lloyd Jeffress's aunt and uncle, who said that it was his duty to himself to get an education worthy of his ability. Finally, in mid-September he received a letter of acceptance from Oregon Agricultural College.

Linus immediately put aside his doubts and gave notice at the machine shop. Mr. Schwietzerhoff was sorry to lose him but told Linus that he would be a success at whatever he chose to do. There would always be a job for him if he came back. Linus filled his diary with resolutions. He would make at least a 95 in mathematical analysis, take as many mathematics courses as possible, make use of his slide rule, and go out for track.

He was particularly concerned about fitting in socially. He knew few college students or graduates, other than his teachers. His cousin Mervyn had attended OAC, but he was in Condon for the summer. Linus sought out advice from vacationing college students in Portland. On the train to Oswego one Sunday he sat next to a former OAC student, who advised him not to take his studies too seriously, to save time to "do something for the school," and to be wary about joining the first fraternity that asked him.

Linus was particularly mystified about fraternities. His Aunt Goldie told him that Mervyn had been asked to join one because he was good in mathematics. Linus was reassured. On that basis, he naively believed, he would be invited readily. His mother accepted his decision to go on to college only because he insisted that he was going anyway. Financially, he could expect no help from her. He spent his last precollege weekend visiting his grandmother, who cried and made him promise to write every week.

If Linus was going to college at sixteen, his mother wanted to make sure of his living arrangements. On Saturday, October 6, 1917, she accompanied him on the electric train to Corvallis, where he was to share a room in a boardinghouse with Mervyn and another young man. He would not see her again until Christmas since there was not enough money for frequent trips home. He had saved two hundred dollars. Tuition was sixteen dollars a semester, board and room twenty-five dollars a month. He planned to work part-time to earn the extra money he would need.

Linus had already met the head of the math department at the train station. The professor asked whether he was new, and Linus told him of his plans to take every course offered in mathematics. He registered for two courses in math, two in chemistry, mechanical drawing, introduction to mining and use of explosives, and modern English prose, as well as military drill and gymnastics. The English course was the last course in the humanities he ever took. He was, however, eager to build up his chest muscles in the gym.

Linus also plunged eagerly into the campus social life. He put on his green cap and joined several hundred other "folks" who serpentined to the football field, where they yelled for OAC, and then on to the girls' dormitory, where they serenaded the residents with "How Green I Am." Twenty sophomores guarded them as they went through the traditional paces. He was invited to a rush party at the Sigma Chi house, but he wasn't invited to join the fraternity. He did join the Miner's Club and was pleased to wear the Miner's pin and be asked to serve on a committee.

He also fell in love, with a coquettish freshman named Irene, who told him frankly that she never permitted any boy to hold her attention very long. He told his diary that he would "show her," but wouldn't try to monopolize her. He spent a bit too liberally, taking her to shows and games, and by October 30, he had used up $150 of the $200 he had saved. The references to girls, movies, ball games, and new acquaintances in his diary end at this point, as did the romance with Irene.

Linus worked a hundred hours a month in the girls' dormitory, chopping wood for the cookstoves, mopping up the kitchen, and cutting up quarters of beef. He was paid only twenty-five cents an hour and his funds kept running short. He quit eating at the boardinghouse and had only one hot meal a day at a cheap restaurant off campus. He listed every expenditure in his diary: two cents for shoestrings, cough drops, or a scratch tablet; ten cents for a sundae, a shoe shine, or eight blue books in which to write his nearly perfect examination papers. He spent eighty dollars during October 1917 but was able to cut expenditures to thirty-seven dollars in November.

In his difficult straits, he turned to his family for small loans. His cousin Mervyn lent him four dollars, which he could ill afford as he also was working his way through. He got a few dollars for Christmas gifts and was able to repay Mervyn. On March 3, 1918, he informed his diary that he had $20.88 left in the bank and $0.25 in his pocket. Fortunately, his mother lent him $5.00 and he was able to find another part-time job to help fill out the school year.

When the year ended, Linus and Mervyn Stephenson spent a month as laborers at a reserve officers training camp at the Presidio in San Francisco. They explored the strange, beautiful city, their first exposure to the world outside Oregon, and had their pictures taken in uniform with two "flapperish" California women. After they left the training camp they earned good money at the shipyards in Tillamook, sailing to work each morning across Tillamook Bay to a three-thousand-ton wooden ship with steel bracings. The world war had stimulated the economy, and jobs were plentiful.

Linus returned to Corvallis in the fall of 1918 feeling infinitely older, wiser, and competent to deal with the demands of college life. He knew there were other schools that would better meet his academic needs, but Corvallis was the cheapest and nearest to home and he knew where to find work around the campus. Lloyd Jeffress had graduated from high school and was Linus's roommate at Corvallis for a year before transferring to Berkeley. Linus was viewed as something of a grind by the other students but was forgiven because of his exuberant love for learning and because he had to work his way through school. He was even asked to join a fraternity.

There was no hint of Pauling's later antiwar convictions at this stage of his life. He simply took it for granted that he should support the American war effort. He took the military training given to male Oregon Agricultural College students, and in May 1918 he was even a runner-up in the contest for best soldier on Military Inspection Day.

The year went quickly, and when he went home for the next summer his mother had a job waiting for him as a milkman. It paid well because it required working from about eight o'clock in the evening to four o'clock in the morning delivering milk to about five hundred customers. Linus enjoyed learning the route during the first week, but the next three weeks were intolerably boring. The horse knew where to go as well as he did. He had promised to keep the job for one month, and he quit promptly on deadline. Belle was quite annoyed, as she still hoped he would remain at home and bring in a regular paycheck.

Linus then found his own job as a paving plant inspector, working for a chemical engineer who had a contract with the state of Oregon to inspect the blacktop pavement on a main highway in southern Oregon. He had previously been rejected as too young, but the engineer could find no one else and finally hired him. The task gave him time to read and do experiments in his spare time. He went through a handbook of chemical facts and tabulated the properties of substances. He tried to find an explanation for magnetic properties. He knew nothing of the published literature on the subject but tried to find the answers on his own by looking for correlations between the values given in the charts. He never solved the problem, however, until he found the answer in the literature as a graduate student five years later.

His theoretical speculations did not prevent him from performing well as a paving inspector, and he was offered regular employment each summer thereafter. With summer employment, he had a secure way to finance his education. Following family custom, he sent his entire salary, $125 a month, to his mother for safe keeping. He had few living expenses since he lived in a tent with the construction workers and had his meals provided.

The end of the summer brought a cataclysmic disappointment. Belle sent word that she had needed money, had used his, and could not return it. He had no money to return to school and had to stay on with the paving firm. To be stranded in the mountains of southern Oregon at the time when his classes were beginning was a bitter disappointment, although his employers were glad to have him. His mother was ill and thought he should be content to be a well-paid chemical engineer and to help care for his sisters. Linus saw no alternative but to swallow his frustrations and anger and wait until next year.

Early in November, however, Linus was delighted to receive an offer of a job as an assistant instructor at Oregon Agricultural College. The war had drained the campus of young instructors while enrollments were up. At eighteen years of age, Linus was asked to teach the sophomore course in quantitative analysis that he had taken the previous year. The salary was one hundred dollars a month. He worked forty hours a week in the laboratory and classroom, which left him no time to take classes although he learned touch typing from the departmental secretary. He did some reading on molecular structure in the library, where he discovered a 1916 paper by Gilbert N. Lewis of Berkeley and some papers by Irving Langmuir. He was much impressed by their work on the electronic structure of molecules and shared electron pair bonds. He had heard nothing of this at Corvallis, where the "hook and eye" theory of the chemical bond was still taught.

When Linus resumed his junior year in the fall of 1920, Oregon Agricultural College had little to offer him academically. He continued to avoid classes in the humanities and social sciences, there were too few physics classes to meet his needs, and he had exhausted the mathematics offerings, which did not go beyond calculus. He looked with envy at his friend Jeffress, who had transferred to the University of California at Berkeley, but OAC continued to offer him financial support. Professor Graf chose him to be his assistant in an advanced mathematics course. The job required a grasp of mathematics and physics at as advanced a level as was taught at OAC.

His first assignment required him to read a set of examination questions on engineering and dynamics that were written on the blackboard, to solve them, to grade the thirty-five students' answers to the problems, and to record the grades in the class book. Professor Graf gave Linus explicit instructions, then left the room, taking the class book with him.

When he returned in twenty or thirty minutes to see what progress the new employee was making at the tedious task set him, Linus was sitting on a high stool, impatiently twiddling his fingers.

"Why aren't you working?" Professor Graf asked with equal impatience.

"You didn't leave the class book. I couldn't finish," Linus answered, grinning.

"But first you have to solve the problems, then grade the papers, and then—"

"I've done all that," Linus said. "I solved the problems in my head, I graded the papers, and I made an alphabetical list of the students' names with the grades noted. If you will now give me the grade book, I shall record them."

Professor Graf insisted on looking over the papers, each of which was corrected and graded appropriately. The professor and assistant arrived at an understanding. Linus was not to work by the hour for a set fee of twenty-five cents, as his predecessors had. He was to be told what work was expected of him, and do it at his own rate at a monthly wage of forty dollars, a mutually beneficial arrangement.

While academics offered little challenge for Linus, he was still somewhat insecure socially. He had entered college at sixteen, not entirely without experience with girls, but too young to interest the coeds in his classes. His early attempts at dating had proved too expensive and troublesome. When Lloyd Jeffress left for Berkeley, however, Linus moved into a fraternity house. His fraternity brothers demanded that he conform to the house rule that every brother prove himself popular by having a date on Saturday night. The penalty for being dateless was to be "dunked" in the bathtub. Linus let too many dateless Saturdays go by, then panicked his fraternity brothers by breathing deeply to oxygenate his blood, conserving the oxygen by not struggling, and remaining supine under the water for a full minute and a half. Having had his fun with the social committee, he arranged to take some amenable student to the movies often enough to avoid further brotherly discipline.

In the winter of his senior year, OAC was short of instructors for over a thousand students who had enrolled in freshman chemistry. There were chemistry classes for chemists, for mining engineers, and for home economics majors. Two women enrolled in the class for mining engineers, but the classes for home economics majors were entirely female. Linus was once again offered a teaching position in addition to his planned work load as an assistant. He also chose three other senior students as teaching assistants.

The first class assigned to Linus had twenty-five freshman home economics students. They were the pert young women of 1922 with short marcelled hair, or with long hair teased into puffs over their ears. They wore short dresses with low waistlines and splashed through puddles in great black galoshes that were always left unbuckled so that they "flapped" as they walked.

Linus was eager to establish himself as an adult authority figure with these young women, who were scarcely a year or two younger than he. Their fall term teacher had been a mature and competent man who had maintained their respect. This teacher, who they were sure must be at least twenty-five since he was already a teacher, seemed dignified and spoke well. His hair was parted neatly in the middle and plastered to his head, but around its fringes little yellow curls kept escaping impishly. He was now a good-looking young man, filled out enough so that his strong features were better proportioned to the rest of him.

Linus took refuge behind his grade book and decided to give them a stiff review of their last semester's work. He chose a name arbitrarily, selecting one that he was sure he could pronounce correctly. He was afraid of setting off an unquenchable storm of giggles by even a slight mispronunciation of a name.

"Miss Miller, will you please tell me what you know about ammonium hydroxide?" Ava Helen Miller responded accurately and precisely, and the class settled down to work. Linus had fallen in love. Fearful of losing the authority of his professorial role, however, he kept his distance from Ava Helen for the rest of the semester. She was exceptionally pretty and often had male visitors at her lab bench although she needed no help. Linus found out a good deal about her, however, from one of his fraternity brothers, who was dating one of her older sisters.

Ava Helen was one of twelve Millers from Oregon City. Her mother was rearing the twelve children alone and could give her no financial support. She was a talented pianist, enjoyed poetry, art, and literature, and was socially and politically sensitive. She was also a quite competent chemistry student, as Linus could observe for himself. He told his fraternity brother, Fred Osborne, how much he liked Ava Helen, and Ozzie found out from his girlfriend that Ava Helen admired and was attracted to her chemistry instructor. Still, teachers were supposed to remain aloof from their students, and one of the other teaching assistants had been reprimanded for showing a romantic interest in a student.

All too soon, spring came and Linus was soon to graduate and leave the campus. There was little time left to approach Ava Helen. One spring day, he handed her corrected notebook back to her. As Ava Helen tells the story, inside she found a note from him that read, "You are to understand that an instructor has been very much criticized for the attention he paid one of his students. I hope this does not happen to me." During the rest of the period, Ava Helen seethed in silence. Did he honestly believe that she was trying to embarrass him by flirting with him? When the last lingering student had left the room she began to scold him. He brushed her comments aside and said, "Wait until I get my things together and I shall walk across the campus with you."

His approach was awkward, but it worked. Their walk across campus was a public demonstration that he was taking an interest in a certain freshman girl, even if she were one of his students, and that his intentions were honorable. She walked quietly beside him as he strode across campus parading his choice for all to see. In a few weeks, they were engaged.

Professor Graf had not kept up with his assistant's personal affairs. He saw Linus at the senior class picnic with a pretty freshman clinging to his arm and Linus introduced her as his bride-to-be.

Dr. Graf asked bluntly, "You are in love with Linus? You want to marry him?"

Ava Helen was positive. "Oh, yes," she said, "and we plan to have lots and lots of children."

Since Linus had supported himself completely since he was sixteen, there was nothing his mother could do about his engagement or about his plan to attend graduate school rather than taking a job after graduation. She had neither the will nor the strength to oppose him. Her wartime marriage to a foreman at a logging camp, whom she had met while he was an impressive soldier in uniform, had soon ended in divorce. Her chronic illness was worsening. Linus gave her little sympathy or attention, partly because he was not fully aware of her medical condition.

His independence was all the more difficult for his mother and other kin to accept because he obviously was enjoying his work, having fun while the more responsible members of the family drudged away at jobs that had to be done. His mother, anxious, overworked, and terminally ill, found it difficult to explain his playfulness to the roomers in her boardinghouse or to her friends and sisters. There were so many jobs for which he could now qualify. He could teach high school chemistry, become a state civil servant, or find a job as an industrial chemist. At graduate school he would earn less than a milkman or mechanic's apprentice and would have to pay living expenses away from home.

The prolific Miller family, which met Linus soon after graduation, accepted him dubiously. The Millers and the Paulings, in private conclaves, each regretted the lack of wealth of the other family. Each agreed that there would be no financial help for Linus and Ava Helen. The Millers were somewhat mollified to discover that Linus was the nephew of Judge James Ulysses Campbell, the distinguished circuit court judge married to Herman Pauling's sister. One of the older Miller girls had taught Judge Campbell's daughter at Oregon City, and he had been a friend of the Millers for many years.

While his family did not support him, the faculty at Oregon Agricultural College were strongly behind him. After all, he graduated with a 94.29 grade point average (despite an F in one gym course he was too busy to attend) and was the class orator at the graduation ceremony. The faculty members had written effusive letters in support of an earlier application for a Rhodes scholarship (which was turned down), and they enthusiastically encouraged Linus to go on to graduate school. Floyd Rowland, head of the department of chemical engineering, and John Fulton, chairman of the chemistry department, both used their contacts to try to secure Linus a place in one of the outstanding graduate schools. Although Rowland was not a distinguished scholar, he valued creativity and talent in his students and made it a point of honor to see that every promising student went to graduate school even if he had to lend him money to do so. Rowland liked playing a paternal role with his students and knew each as an individual.

Linus very much appreciated the supportive atmosphere at OAC and the advice of sympathetic faculty members such as Rowland. OAC valued him as an instructor and as one of their most promising students, and let him work at his own pace. He took whatever courses he wanted and read the journals and did experiments on his own. He wanted a similar environment for graduate school. Harvard University accepted him, but insisted that as a half-time instructor he should take five years to work on his doctoral degree. Linus didn't think it would take him that long and looked for an institution that would be more flexible.

He was inclined to go to the University of California at Berkeley, which had an excellent chemistry department. Lloyd Jeffress was there, majoring in psychology. But Berkeley had not responded to his application when he got an offer from the new California Institute of Technology at Pasadena. Caltech was in its fourth year under its new name (it had previously been called Throop College) and was eager to attract promising students. Pauling had corresponded with the chairman of the chemistry department, A. A. Noyes, whose department offered him a fellowship. Pauling went down to Pasadena and received a warm reception from Noyes, who was pleased with his longtime interest in crystals and suggested that he read a book on crystallography by W. H. and W. L. Bragg during the summer. Linus accepted the offer of a fellowship and went to considerable trouble to obtain the book, which he finally located at the Oregon State Library in Salem. He took it with him when he left Ava Helen's home and went to his customary work stint as a paving inspector, this time at Warrenton on the Oregon coast.

The importance that Linus placed on finding this particular book mystified his mother, sisters, aunts, and future in-laws. A book on rocks? The library was full of books about rocks. Linus made no effort to explain that he was interested in understanding the links between the atoms that formed the molecules within the rocks. Ever since he had read an article by Gilbert N. Lewis that suggested that a bond between two elements could be formed not only through the transfer of electrons but through the sharing of one pair of electrons, he had been bubbling with curiosity about how the complex linkages between atoms could be deciphered. None of this could be shared with his mother and sisters.

His fiancée, at least, had confidence in him and his plans. She had first seen him as a brilliant college teacher; he had seen her as one of the best students in his class. They had a meeting of the minds about their shared future. He was to be a professor of chemistry on the campus of some distinguished institution. She would be his wife, rear his children, and encourage him in his research and writing. They were not about to let the doubts and nagging of his mother or the critical teasing of the boarders change their plans.

3

Hidden Patterns, 1922–1925

When Pauling went to the California Institute of Technology in the fall of 1922, he had no idea that he would spend forty-one years there. Had the University of California at Berkeley answered his application more quickly, he might have gone there instead—a move that probably would have made little difference to Pauling, but would have radically altered the future of chemistry at Caltech! Pauling didn't wait to compare offers, but simply accepted the first offer of admission that was extended to him. He entered Caltech at a time when it was just beginning the period of dramatic growth and expansion that would convert it from a small manual training school (Throop College) to one of the world's leading technical universities.

The course work at Caltech was much more advanced than that at OAC, but Pauling did not find it difficult. In his first semester, he was particularly influenced by Professor Richard Tolman's course Introduction to Mathematical Physics, which gave him the basic principles he needed. He took almost all the important chemistry courses during his first year as a graduate student, as well as many in mathematics and physics, including advanced algebra, higher dynamics, thermodynamics, chemical thermodynamics, vector analysis, Newtonian potential theory, quantum theory, physical optics, functions of a complex variable, and integral equations. Pauling modestly observed that these courses "helped me to overcome the handicap of a lack of knowledge of physics and mathematics."[1]

Pauling took as many graduate courses in physics during his career as did the candidates for doctorates in physics. Since there were no rigid requirements, he often registered for sixty hours of classes and twenty hours of research. The new school did not yet have rules limiting the number of courses a student could take.

Even with all his classes, Pauling found time for extra study and seminars beyond the required work. He attended weekly joint seminars sponsored by the Mt. Wilson Observatory and Caltech, and lectures by visiting foreign scientists such as Niels Bohr, Arthur Sommerfeld, Albert Einstein, Peter Debye, and Paul Ehrenfest. He also attended, for three years, a weekly seminar in physical chemistry presided over by Professor Tolman, who used the seminar to keep track of the progress of the students. One day he asked Pauling to describe magnetism and to explain its source.

Linus replied frankly, "I don't know. I haven't had that yet." Later one of the other students, Richard Bozorth, remonstrated him, "You know, Linus, you shouldn't have said that."

"Said what?" Pauling was bewildered.

"You shouldn't have said, 'I haven't had that yet.' Now that you are a graduate student you are supposed to know everything." This idea hit Pauling with a tremendous wallop and he resolved never again to be caught ignorant of fundamental chemical facts. Caltech was a perfect place to fill in the few remaining gaps in his already prodigious knowledge.

The only thing missing in his life at this point was Ava Helen. She was still a home economics major at Corvallis, and both found their continued separation intolerable. He was not able to afford to go home for the holidays during the school year, so they made plans by letter to be married in Salem on June 1, 1923, at the home of the bride's older sister, Mrs. Walter Spaulding. Linus bought a 1916 Ford for fifty dollars from one of his professors and headed north for Oregon as soon as classes were out. In his haste he attempted to drive through the night, became overtired, and ran the car into an excavation in the hillside along the side of the road where a construction crew had been working. He was able to maneuver out of it, however, and arrived on time with only a battered fender.

The families put aside their doubts about the marriage and turned up in good spirits for the wedding, as did Lloyd Jeffress and his aunt and uncle. The account that they supplied to the Salem newspaper mentioned that the bridegroom was "working in a laboratory in Pasadena." To admit that he was still a student would have meant a loss of face.

After the wedding, Linus and Ava Helen honeymooned briefly at the home of some vacationing friends in Corvallis. They then stayed in the Pauling home in Portland until Linus could be set up for his summer job as paving inspector. This year he worked for the Warren Construction Company, which sent the couple off to eastern Washington in a Pullman sleeper. He completed a summer's work in eastern Washington and Oregon and returned to Pasadena in the fall with his bride.

Pauling entered graduate school with a driving urge to do research, and he started doing interesting research almost immediately on arrival at Caltech. To understand Pauling's research work, and why it was important, a little background in the history of physical chemistry is required. Physical chemistry was, in Pauling's student days, a fairly obscure field of study. But it is now one of the most fundamental areas for research in chemistry. It relates to basic laws of matter that allow the formation of complex structures—such as crystals, molecules, chemical compounds, and, ultimately, organisms such as human beings. These laws also describe how atoms rearrange themselves spatially from one set of positions to another. They tell us how *chemical reactions* convert one set of substances into another, as when gasoline and oxygen burn to form carbon dioxide, carbon monoxide, and water.

Chemistry is the study of combinations: the study of the microscopic elements of matter and the way these elements combine to make macroscopic substances. Physics, most generally conceived, is the study of the physical world, but in practice physicists generally concern themselves with the study of relatively simple physical systems that can be modeled by tractable mathematical equations. Physical chemistry lies at the boundary of the two disciplines. The forces inside the atom and molecule are the domain of physics; the behavior of molecules as they form new molecules is the domain of chemistry. And the behavior of atoms, as they combine to form molecules, lies at the intersection of physics and chemistry: it is the essence of physical chemistry. Pauling entered graduate school at a very exciting time in physical chemistry, because physics itself was in the process of revolution. No one knew exactly how this revolution was going to filter up to the world of chemistry—but it was clear that this was going to happen, one way or another.

One might say that what computers and molecular biology are today, physics and physical chemistry were then: thrilling new frontiers of research, in which theoretical insights and experimental breakthroughs followed one another in rapid succession, so quickly that only the experts were able to keep up. The older, "classical" views of the physical world had been conclusively proved inadequate, and scientists from all different areas were working furiously to try to understand what would finally replace them.

In his book on the origins of quantum physics, Friedrich Hund describes the history of physics as "a gradual realization that all is not just as it seems—a departure from the 'perceptual'." A vivid early example of this is the transition from the physics of Aristotle to that of Newton. Aristotle's physics held that the force experienced by a moving body was proportional to its velocity. This is a simple, intuitive idea, which matches up with the everyday experience of dragging a load. Pulling something faster requires more force; greater speeds meet greater resistance. But in the seventeenth century, Newton broke from this idea and declared, counterintuitively, that force is proportional to *acceleration,* the rate of change of velocity. Of course, Newton was right—but the catch is, he was only obviously right for the case of a particle moving through a perfect vacuum. To understand the real world in terms of Newton's physics one must introduce the extra assumption of *friction,* which is resistance proportional to velocity. Aristotle's theory is simpler, and at first glance seems to explain our daily "perceptual" experience better. But on careful consideration of all the data, Newton's theory can be seen to be far superior; the slight burden of added complexity is far outweighed by the increased accuracy.

In general, the result of this kind of departure from the perceptual is that matters that previously seemed simple now must be treated as complex. Aristotle just had force; Newton needed force plus friction. Or, to take a more relevant example: to the Greeks, earth, air, fire, and water were absolutely simple; they were the indivisible elements from which everything was constructed. But in the light of modern chemistry each of these "elements" is seen as a complex entity composed of numerous chemicals interacting in various ways; indeed none of them is yet fully understood.

The beginning of the twentieth century was a period of unprecedented activity for this process of fundamental scientific advance. Just as Newtonian physics had supplanted the more intuitive ideas of Aristotelian physics, Newtonian physics itself was about to be superseded by the far less intuitive ideas of quantum theory and relativity.

Pauling's work and interests would not require him to become truly fluent in relativity, but they would require that he understand the quantum revolution. Quantum physics was redefining the nature of the atom and as a physical chemist Pauling would spend his life figuring out how these newly defined atoms combined into large, complex molecules.

Quantum physics began rather unromantically, in the year 1900, with Max Planck's attempt to explain the peculiar properties of *black-body radiation,* that is, why light radiating out of a pin hole on the side of a box heated to above 1000° C reached only a certain frequency before it peaked and dropped down. To solve the problem Planck made an assumption that energy is released discretely in chunks of a specific size, rather than continuously. These chunks were called *quanta.*

The person who took quantum mechanics the next step was Albert Einstein. In 1905—the same year that he proposed the special theory of relativity—Einstein published a paper hypothesizing the existence of light quanta. His theory also suggested the phenomena that take place in *photoelectric cells,* cells that change light into electricity and are used to detect motion when someone or something walks through a beam. Now used on elevator doors, before 1905 they were just a curious, unexplained phenomenon.

The theory of light quanta also introduced a fundamental concept into quantum physics: the concept of wave/particle duality. Although light could be conceived as a particle that releases energy in finite quanta, that model didn't preclude its also being a wave. Later, scientists would realize that not only light but all particles—for example, electrons—have this capability of being both particle and wave. From this, scientists realized that "wave" and "particle" would come to be seen as just two different ways of thinking about the same underlying reality. But, although these early steps were extremely important, quantum physics did not really come into its own until physicists used it to explain the atom.

The concept of the atom is conventionally traced back to the ancient Greek Democritus. But Democritus was not a scientist; he had no empirical support for his ideas—he was simply speculating that, instead of being infinitely detailed, all matter was ultimately constructed of tiny fragments of finite size, which he called *atoms.* Atomism was not widely accepted by the Greeks, and for centuries it remained a relatively unpopular alternative to the Aristotelian theory that all matter consisted of one of four elements—earth, air, fire, and water. Not until the seventeenth century was the idea of the atom really accepted.

Over time, certain laws of chemical combination were observed. It was noted that when two or more elements combine to form a compound, they always do so in the same proportions (actually this is not exactly true, but it is close enough in most cases to be a useful "law"). For instance, water is composed of hydrogen and oxygen, and, judged by weight, it always contains these two elements in the proportions 1:8. This is called the *law of constant proportions.*

Furthermore, suppose two elements combine to form two or more different compounds. Then there must be a whole-number ratio between the amounts of the one element that combine, in the different compounds, with the same amount of the other. Again consider hydrogen and oxygen, which can form either water (H_2O) or hydrogen peroxide (H_2O_2). The relative weights of hydrogen and oxygen in water are always 1:8, and the relative weights of hydrogen and oxygen in hydrogen peroxide are always 1:16, so that the respective weights of oxygen corresponding with the same weight of hydrogen are given by the ratio 1:2. This is called the *law of multiple proportions;* it explains why there are twice as many oxygen atoms in a molecule of hydrogen peroxide as in a molecule of water.

These "whole-number laws" inspired English scientist John Dalton's 1808 idea that elements are composed of atoms, and that the atoms of each element all have the same weight. Furthermore, Dalton proposed, chemical compounds were formed by a union of definite numbers of atoms of each of the elements concerned. This was the birth of chemistry as we understand it today, in which the problem of the structure of matter is understood as the problem of the arrangements of atoms.

One after another, more laws of chemistry were discovered. In 1811, there was *Avogadro's law,* a key principle that states that a certain volume of a gas contains a certain number of molecules, no matter what element or elements make up the molecule (as long as pressure and temperature do not change). And at about 1860, the concept of valence, or "combining power," began to assume a central role in chemical theory. The *valence* of an atom indicates how many links it can form with other atoms. For instance, a carbon atom always has four ways to link to other atoms. Hydrogen has one link, and oxygen has two. Each atom in a multiatom structure uses as many bonds as indicated by its valence number. Because of this fact, sometimes double bonds at the same site are needed, or even triple bonds.

The idea of valence is not always so easily applied—for instance, iron can have a valence of either two or three, depending on the compound in which it occurs. And substances such as salts require a more sophisticated notion of valence than was prevalent in the nineteenth century. But the basic idea is powerful and widely applicable. Using valence as a guide, similarities were discovered among various atoms, and on this basis in 1868–69 Grigory Mendeleev and others constructed the periodic table of the elements, solidifying the theory of atomic structure.

Another important discovery of nineteenth-century chemistry was the distinction between ionic and covalent chemical bonds. Today it is known that there is no rigid distinction between the two types of bond: some bonds are more ionic in nature, some more covalent, and some thoroughly half-and-half. Quantum mechanics complicates matters tremendously. But the distinction is still in many cases a useful one.

An *ionic* bond is one in which two atoms are held together by electricity: one has a positive charge, the other a negative, and the two thus attract each other. An excellent example is salt, sodium chloride. If one dissolves salt in a container of water and places a positively charged electrode and a negatively charged electrode at different ends of the container, one finds that negatively charged chlorine ions move toward the positive electrode and escape the container in the form of chlorine gas. The positive sodium ions tend not to reach the negative side, as the hydrogen atoms in the water tend to take up all the available electrons. But the point is that the bond clearly involves electricity; thus the formula for salt is often written Na^+Cl^-, instead of just NaCl.

Covalent bonding, on the other hand, is based not on electricity but rather on the sharing of electrons by different atoms. Heisenberg's uncertainty principle, a key law of quantum physics, states that one cannot measure the exact position and the exact momentum of a particle at the same time. A consequence of this is that electrons cannot be precisely localized; instead the electrons in an atom are understood to surround the nucleus in diffuse *electron clouds.* The space in which an electron is likely to be found is called its *atomic orbital.*

The spatial arrangement of the electrons in an atom, according to quantum physics, is characterized by three sets of numbers: the principal, azimuthal, and magnetic quantum numbers. The peculiarity of quantum physics as opposed to classical physics is revealed by the fact that all these numbers are integers. The energy levels in an atom are arranged roughly into main levels or "shells"; the principal quantum number associated with a shell determines the energy of the shell and the size of the orbitals of the electrons in the shell. These shells are denoted by letters; the "k," "l," and "m" shells correspond to principal quantum numbers of 1, 2, and 3, respectively. Each of these shells is in turn composed of one or more subshells, each of which is specified by the azimuthal quantum number; the "s" and "p" subshells correspond to azimuthal quantum numbers of 0 and 1, respectively. Finally, each subshell is made up of one or more orbitals. An orbital within a particular subshell is identified by its magnetic quantum number, which determines its spatial position relative to the other orbitals (the name *magnetic* derives from the use of these numbers to explain the spectra produced when atoms are caused to emit light while in a magnetic field).

The easiest example to understand is hydrogen. Each hydrogen atom has one electron, though it has room for two in its outermost electron shell. It can achieve this desired state, or something very close to it, by sharing its electron with another hydrogen atom, and in exchange gain the capacity to share the single electron belonging to the other hydrogen atom. Another example, to be discussed a little later, is the bond between carbon and hydrogen in methane.

We have said that there is no rigid division between the two types of bond; it is instructive to observe that, with the benefit of quantum mechanics, even the hydrogen-hydrogen bond can be viewed as ionic. If one hydrogen atom gave its electron to the other, we would have H^+H^-; if the second atom gave its electron to the other, we would have H^-H^+. Thus one can envision the covalent bond as a kind of rapid switching back and forth between these two conditions. What Pauling's theory of the chemical bond shows is that this view is actually correct: the covalent hydrogen-hydrogen bond may be viewed as a kind of oscillating superposition, or *quantum resonance,* between these two states; in effect, the pattern is a statistical combination of the two patterns.

Chemical knowledge was advancing at an impressive rate. As to what went on inside the atom, however, the nineteenth-century chemists were essentially ignorant. Inorganic chemists tended to explain valence in terms of electric charges on the atoms; organic chemists generally preferred to think of forces emitting from atoms at particular valence points. None of these models was generally accepted, and for this reason many respected physicists, including Max Planck and Ernst Mach, avoided the idea of "atoms" altogether. But as the turn of the century approached, experimental physics advanced to the point where properties of atoms could be directly measured, and the rough chemical models were gradually replaced by more accurate physical models.

The physicists had established that the atom consisted of a positively charged part, and a number of electrons, particles carrying negative electrical charges. But very little was known about the positively charged part of the atom. A few years after 1900, however, it was realized that the distinguishing property of any atom of any chemical element was the number of positive charges it contained. And then Ernest Rutherford, in a series of clever experiments, concluded that the positive charge and most of the mass of an atom must be concentrated in its center.

Rutherford's model of a positively charged nucleus composed of protons at the center of every atom surrounded by rings of rapidly spinning electrons created a dilemma for scientists. If electrons were really rapidly spinning around a positively charged proton nucleus, they would, according to classical ideas, quickly use up their energy and spiral down into the nucleus. What prevented them from doing so?

It would fall to the great physicist Niels Bohr to answer the question. In 1913, he published three seminal papers combining early quantum theory with the Rutherford model and the theory of *spectra,* that is, the theory that atoms of any element emit or absorb light at only certain specific frequencies. Scientists had tabulated these frequencies for many elements. These papers of Bohr's constitute the first detailed theory of the structure of the atom and provided plausible arguments as to why this model of atomic structure would account for stable atomic structures.

Bohr saw that the stability of the atom—the fact that the atom's negative electrons don't collapse onto the positive nucleus—could be explained by the fact that these rings or orbits were themselves "quantized." They could exist only at certain fixed radii from the nucleus—not in a continuous range of distance, as in classical physics. Electrons could jump from a lower ring to the next higher ring only when excited to a specific level. Electrons also returned to lower levels only by releasing that exact "quanta" of energy that had previously excited them to a higher level.

In this way the familiar "solar system" model of the atom, as shown in figure 1, was conceived: a positively charged center or nucleus, surrounded by electrons in rings of increasing diameter. Of course this "solar system" diagram is an oversimplification, a mere heuristic device for explaining a mathematical theory. It is almost a caricature—a great deal more was known about electrons than such a picture can communicate. The nature of the nucleus remained at this stage largely a mystery. Most important for physical chemists, however, was Bohr's realization that the chemical properties of atoms were determined by the number of rings of electrons they had, and that the outermost ring of electrons was the most crucial.

Not everyone accepted Bohr's ideas: in fact, the outstanding physicists Otto Stern and Max von Laue promised each other that if there were "anything in this nonsense of Bohr's," they would give up physics for good.[2] But there was indeed something in Bohr's theory—and, fortunately, Stern and von Laue reneged on their promise and continued to do outstanding work in physics. By 1918 Bohr's sometimes vague ideas had inspired a generation of physicists, and the foundations of what is now called the "old quantum theory" were complete. More and more fine details were gradually filled in. For instance, Bohr's model had not explained a number of aspects of atomic structure, such as the fact that the maximum number of electrons in each ring was always 2 for the first, 8 for the second, 18 for the third, and 32 for the fourth ring. But later research within the same theoretical framework provided answers.

As the description suggests, the old quantum theory was not entirely adequate. It suffered from what many considered internal inconsistencies—it was neither entirely a quantum theory nor entirely a classical

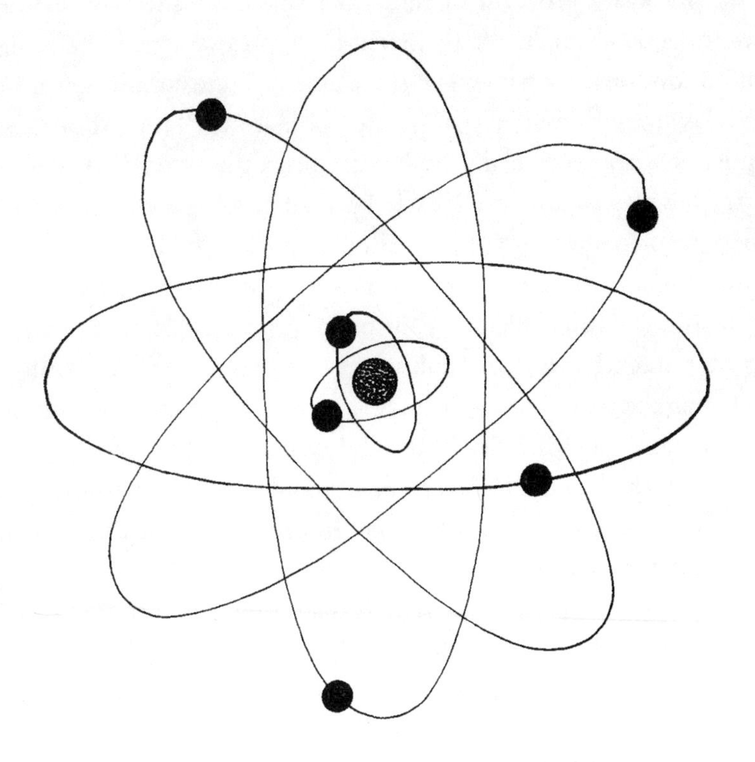

Figure 1. Standard depiction of the Rutherford-Bohr model of the atom.

theory. Most importantly, it failed to predict the states of the helium atom or any other atom but hydrogen. It seemed to be incapable of dealing with the interactions between electrons. To address these problems, significant changes were required, and the mid-1920s saw another round of fundamental advances, most notably Werner Heisenberg's matrix mechanics, Erwin Schroedinger's wave equation, and the work of Paul Dirac and Max Born. With these advances the old quantum mechanics gave way to the new. The new quantum physics provided a mathematical and conceptual framework that explained puzzling features of the old quantum physics, such as the observation that subatomic particles, such as electrons, could sometimes appear as waves and sometimes as particles.

Pauling entered the picture at a time when physical chemists, excited and confused by the new quantum physics, were trying to figure out the various ways in which atoms of the same element might bond to each other. This was a problem of fundamental importance, since many of the characteristics of materials depend not on the atoms that compose them but on the arrangement of these atoms. For example, graphite and diamonds are both made up of pure carbon atoms. The carbon atoms in graphite are arranged in flat hexagonal rings that are fitted together in layers. These layers are only loosely bonded and break apart readily; this property explains why graphite is slippery and useful as a lubricant. The carbon atoms in diamonds are arranged in a three-dimensional structure, with each atom linked to four other atoms in different planes. This makes for an extremely hard substance that can be used in a drill point. Other characteristics of molecules, such as their color, transparency, ability to conduct electricity, and stability, can also be attributed to their structure when they are joined to form molecules. The pattern in which the atoms fit together is in many ways more significant than the nature of the atoms themselves.

Even today, figuring out the structure of molecules is not an easy task. The sizes involved are so small—the diameter of an atom is about 10^{-8} centimeters—that direct observation is not possible. Now we have high-tech experimental tools, such as electron diffraction, and we have high-speed computers that are able to test out a large number of alternative models and, given appropriate criteria, select the best one. But in the 1920s conditions were much more primitive. The state of the art then was a new technology called *X-ray diffraction.*

At OCA, Pauling's mentors had not been particularly well acquainted with the latest findings of quantum physics and the latest laboratory techniques. At Caltech the situation was entirely different. Pauling was assigned to a young professor, Roscoe Gilkey Dickinson. He was Dickinson's only advisee, and although Dickinson was only ten years his senior, Pauling regarded him as a wise elder scientist. Dickinson was familiar with X-ray diffraction, which had very recently been introduced, and he was more than willing to put a promising new graduate student to work on problems at the cutting edge of research.

As an undergraduate, Pauling had been taught chemistry largely as a large collection of disconnected facts. His undergraduate work had proved that he was good at learning facts, and even at discovering new information in the laboratory. But he wanted to uncover the general principles that accounted for the factual observations. At the time Pauling was studying chemistry, mathematics, and physics, trying to broaden his scope of knowledge, he threw himself wholeheartedly into study of X-ray diffraction.

Diffraction itself was not a new phenomenon; it was known to Leonardo da Vinci. Augustin Fresnel, a nineteenth-century French physicist, gave the first explanation of diffraction in terms of the wave theory of light. The crucial idea of diffraction is shown in figure 2, for the special but important case of diffraction by the edges of a disk. Light, assumed to emanate from a point source, travels through space as a series of waves, which may be visualized as a concentric series of crests and troughs, much like circular ripples in a pond. A disk that corresponds in size to the wavelength of the light source being used is held up between the light source and the observer's eye, as shown in the figure. The edges of the disk scatter the light, thus reradiating the electromagnetic energy. Each part of the circumference of the disk acts as a new source of light, and new waves radiate from it. The waves formed in this way recombine and interfere with each other. Sometimes a crest from one point and a trough from another meet, and cancel out. Sometimes two crests meet and form a larger crest; sometimes two troughs meet and form a deeper trough. Reinforcement tends to occur in certain specific directions, which are determined by the wavelength of the light and the size of the disk. The straight lines on figure 2 show the mathematical functions that describe the intersection points. The curved lines are the waves of light.

Scientists get different kinds of diffraction patterns by shining light through gratings with equally spaced, narrow lines. With this kind of diffraction grating, the several lines act in concert, since diffraction from all lines occurs at once. And when one uses two gratings perpendicular to each other, as illustrated in figure 3, the scattering occurs only at the

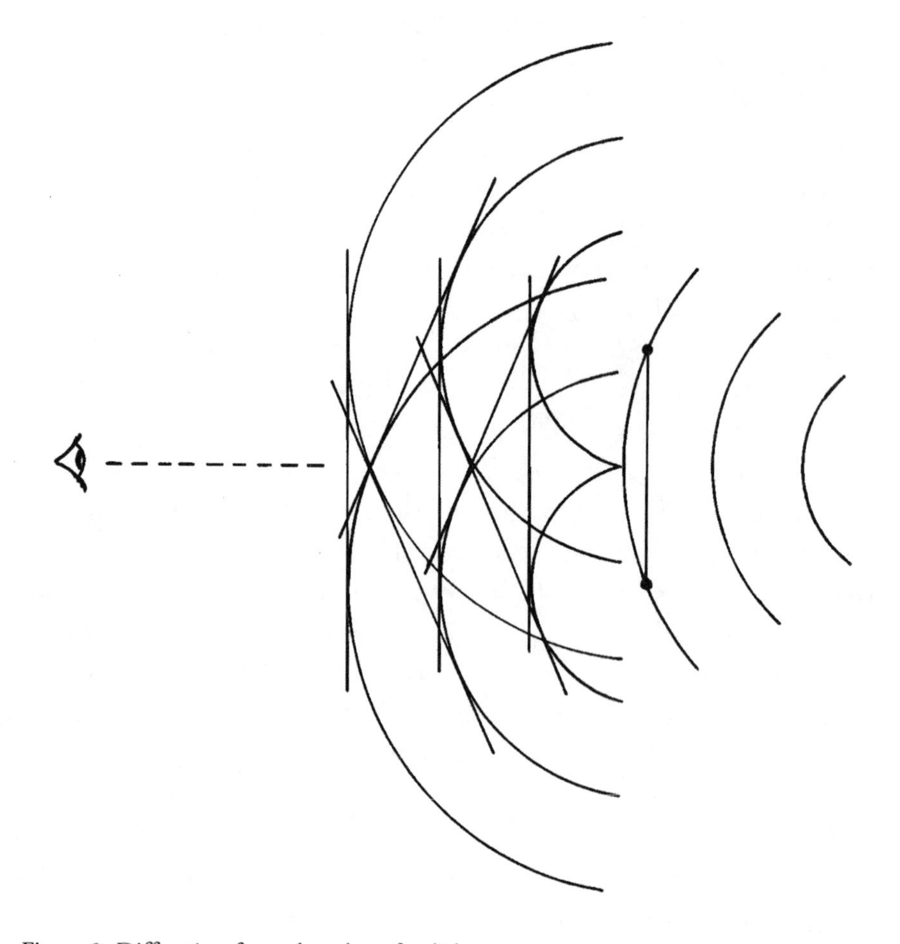

Figure 2. Diffraction from the edge of a disk.

intersections between the two lines. Another way to do this is to puncture a piece of tin foil with an array of pinholes, and shine a monochromatic light through it from a single point.

To understand the use of diffraction in chemistry, one must imagine an invisible object, that one cannot hear, see, smell, taste, or touch. The only way to study this object is to shine light through it, or reflect light off it, and observe the diffraction patterns that result. By observing the diffraction patterns formed by the light, and matching them up with patterns caused by different kinds of diffraction gratings, one can obtain a great deal of information about the structure of the object. This is not an easy task—it requires one to "reason backward" from the diffraction pattern to the structure that produced it. But this inference problem is not insurmountable, especially if one has a little background information about the structures involved.

This is precisely the situation faced by physical chemists. The objects involved—molecules—are not invisible, but they are so small as to be undetectable by our senses, even when amplified by devices such as microscopes. The catch is that ordinary light is not sufficient to produce diffraction patterns by passing through molecules—the wavelength of ordinary light is too large, compared to the size of the molecules. What is needed is a kind of light with a tiny wavelength, a wavelength of molecular dimension. This requirement, as discovered by von Laue and his students, is fulfilled spectacularly by X-rays. British scientists W. H. Bragg

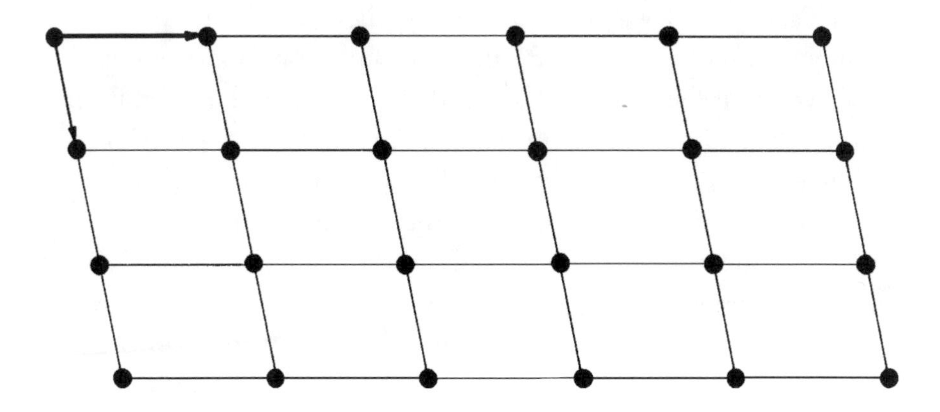

Figure 3. Two mutually perpendicular diffraction gratings.

and W. L. Bragg determined the diamond and zinc blend structures among the first done by X-ray diffraction.

While the first applications of X-ray diffraction were to the study of crystals, Pauling and other scientists would apply it to the study of proteins and other biochemical molecules. But when Pauling began graduate school, X-ray diffraction was still restricted to simple systems such as salts because of the unusually regular, repetitive structure of salt crystals. A crystal is not quite so regular as a diffraction grating, but it has a repetitive geometric structure that makes the interpretation of diffraction patterns a relatively tractable problem. For a simple example, consider the familiar substance sodium chloride, or salt. Salt normally occurs in a repetitive cubic lattice, each cell of which is an interpenetration of two face-centered cubes, one for sodium and one for chlorine, as shown in figure 4. If we draw lines between the atoms in the center of the crystal, we can see that they form a tetrahedron. Each of these two component lattices would seem to produce the same diffraction pattern, but there is one important difference: the diffracted beams are not in phase. Thus there is a superposition of two diffraction patterns, which can be separated easily if one knows the crystal structure. If the crystal structure is not known, the separation can still be performed, but it is not such an easy matter.

Pauling, under Roscoe Dickinson's guidance, started to work immediately on a synthesis of the substance lithium hydride. Dickinson's conjecture was that the hydrogen atoms in this crystal have negative electric charges, each hydrogen atom having taken an electron from a lithium atom. Pauling's task was to check this hypothesis using X-ray diffraction. Without the tool of X-ray diffraction this work would have been impossible; with it, it was merely difficult. After three weeks, however, Pauling learned that this same experiment had been performed successfully by two researchers in the Netherlands. Thenceforth, he read the scientific literature more carefully before initiating an experiment.

Pauling spent another frustrating month making crystals of fifteen inorganic substances and subjecting several of them, without success, to the first stages of X-ray diffraction investigation. At this point he was grateful for the help of Dickinson, who carried him patiently through the various steps in determining the complete structure of a selected crystal, the mineral molybdenite, MoS_2. Dickinson was a clear-headed

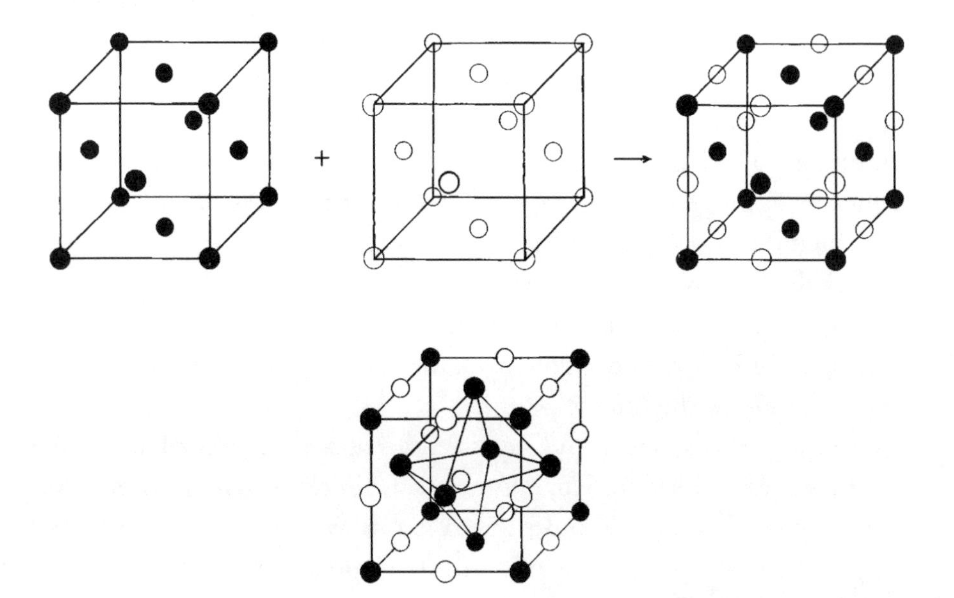

Figure 4. Crystal structure for salt: two superimposed face-centered cubic structures.

and thoughtful young man. The process of structure determination involved a succession of logical arguments that were presented by Dickinson in a meticulous way with an emphasis on rigor. He was highly critical of carelessness and superficiality.

Pauling and Dickinson's study of molybdenite was successful. The molybdenum atom was found to be surrounded by six sulfur atoms at the corner of a trigonal prism, rather than an octahedron, as might have been anticipated from earlier structure determinations. (A trigonal prism is the shape of a typical dimestore prism, the surface swept out by lifting a flat triangle a certain distance off the page.) This shape had not previously been observed to coordinate the structure of metal atoms. This was a small triumph for both of them.

Having mastered the technique of handling the laboratory equipment and been rewarded by an exceptional finding, Pauling worked with tremendous enthusiasm. But technical competence is not necessarily rewarded with dramatic achievement. Weeks passed, more weeks passed, and there was little of interest to record in the way of findings.

Pauling stopped working in the laboratory and took time to write a carefully detailed paper describing the work on molybdenite. When the paper was done, he signed it with his own signature in his neat spare script as he had always signed his papers in college and in high school. Dickinson immediately recognized that the paper was publishable and resented the fact that he was not listed as coauthor. He spoke to Noyes, who called Linus a few days later.

After Noyes had praised the paper and talked casually about its contents, he said, "You know, don't you, that Dickinson made a rather large contribution to this study?"

Pauling readily agreed and expressed his appreciation for Dickinson's assistance and direction. Noyes continued, "Dickinson built the apparatus, selected the crystal, and showed you how to carry out the procedures you developed, did he not?" Pauling concurred.

"Wasn't it Dickinson who showed you how to carry out the procedures you developed?" Pauling nodded his agreement. "Don't you think," Noyes continued, "that it would be proper for you to ask Dickinson to be named coauthor of this paper when it is published?"

Pauling concurred. He had not intended to publish the paper under his sole authorship, and the question of dual authorship was resolved with a minimum of hurt feelings. In the future, when he was the senior professor, he was careful to assign positions in the list of authors according to his judgment of individuals' contributions. Sometimes he believed that the fact that he had originated a work justified placing his name first, a practice that occasionally led his collaborators, who had done the actual laboratory work, to feel that their contributions were undervalued. In cases where there are more than three authors, the convention is to cite the papers as "Pauling, et al.," which left some collaborators feeling "lost in the et al.'s."

In his second year of graduate study, Linus was eager to establish himself as a producing scholar. He scanned the published literature, looking for errors he might correct. In one case he discovered that some unusual X-ray reflections, which another researcher had reported as occurring on spectral photographs, were nothing more than diffraction from a powder produced by grinding on the surface of the crystal. Linus published frequently during this period, largely critiques of previously published works. He enjoyed this kind of critical work so much that it became a lifelong avocation. In his spare moments, he loved nothing more than poring through the latest scientific journals. Whenever he found an error, he sent off a brief note to the editor to correct the scientific record.

At this stage in his career, he was not yet confident enough to take on the major figures in the field, or to attempt a paper on an important theoretical topic. On one occasion Arnold Sommerfeld, an eminent German physicist, spoke on campus. Sommerfeld presented his theory of the fine structure of X-ray spectral lines, as related to the distribution of electrons among different "shells" in the atom. The *eccentricity of the orbit* of an electron—a measure of how far the orbit deviates from being a circle—was said to be determined by the angular momentum of the electron, which in turn was given by the *azimuthal quantum number.* Sommerfeld's theory was intricate and subtle; the structure of the shells was supposed to be explained by special relativity theory, which implied that electrons with orbits of different eccentricities would have different masses. But the theory also contained some peculiar ad hoc assumptions, such as the unexplained assignment of two different azimuthal quantum numbers, "inner" and "outer" numbers, to each of the electron orbits. These double quantum numbers were a serious conceptual problem, and Robert Andrews Millikan and Bowen later wrote a paper on the topic, "A Great Problem in the Theory of Spectra." Eventually these difficulties were resolved by the Dutch scientists George Uhlenbeck and Samuel Goudsmit using the theory of electron spin, a remarkable and counterintuitive addition to quantum theory according to which each electron has a "spin" of either ½ or –½. The spinning of electrons is not known to be a real physical phenomenon (after all, quantum mechanics does not describe a detailed trajectory for an electron), but it provides a natural interpretation of this otherwise quite mysterious two-valued quantum number. Spin provides each electron with a fourth quantum number, in addition to the three associated with its position in space; spin is thus crucial to the Pauli exclusion principle, which states that no two electrons in any one atom may have the same four quantum numbers. The exclusion principle effectively limits the number of electrons in any one orbital to two and requires that the spins of these two be opposite each other.

It is probable that Sommerfeld himself was aware of the problems with his theory, and that he called these problems to the attention of his students—including Werner Heisenberg, whose own matrix mechanics was soon to supersede Sommerfeld's ideas. But, in his lecture, Sommerfeld presented his theory with such confidence that Pauling never stopped to question how there could be two different numbers. Pauling was later chagrined by this oversight, and he learned quickly that he should follow his own doubts and questions, even when they contradicted the statements of respected senior scientists. He learned that "it was possible for people to make mistakes, or that it was possible to add something to what other people had said."[3]

Linus also continued to work productively in the laboratory, studying the structure of crystals with X-ray diffraction. When it was time to submit his doctoral thesis, so many of the sections of it had already been published that he had little to do but paste the papers together in the proper order with appropriate transitional paragraphs. His brilliance as a young doctoral candidate was widely recognized, and he was eagerly sought after by the University of California at Berkeley and California Institute of Technology, both of which wanted him to join their staff. It is customary for new Ph.D.'s to seek employment in a university other than the one that awarded their doctorate, and Linus naturally looked at other opportunities. While still completing his last year at Caltech, he went up to Berkeley with Ava Helen for a long weekend to visit his friend Lloyd Jeffress.

While in Berkeley, it was natural for Pauling to drop in to see Gilbert N. Lewis, who immediately urged him to become a National Research Fellow at Berkeley as soon as he finished his doctoral work. A. A. Noyes at Caltech was not so happy with the prospect of losing his most brilliant student. The up-and-coming new campus could not afford to give up its best students to more established institutions. He urged Linus to stay for a few months to finish the work he had been doing. He could always go up to Berkeley later in the year, Noyes suggested. Noyes had been something of a father figure to Linus, and it was difficult to turn him down. The three months stretched out to seven, as Linus became more and more immersed in his work.

Later in the year, Noyes thought of another delaying technique. He suggested that Linus apply for a Guggenheim Fellowship to visit Europe and meet the scientists who were delving into the structure of the atom. He assured Linus that he would get the fellowship, supplemented it with one thousand dollars of Caltech's money, and promised to advance him more funds if he should run short on the trip. Pauling was persuaded and wrote a letter resigning the fellowship to Berkeley.

Pauling was not aware, at the time, that he was the subject of a tug-of-war between Noyes and Noyes's former graduate student from his MIT days, Gilbert N. Lewis. Noyes was so eager to keep Pauling on the Pasadena campus that he hurried him off for a Guggenheim travel year, even without the written promise of a fellowship, in order to keep him away from Berkeley. Lewis seldom left the Berkeley campus and hated doing so, but he traveled to Pasadena to remonstrate with Noyes. Noyes talked him out of offering Pauling a professorship as he had intended. Lewis was simply not ruthless enough to steal a favorite student from his own favorite former professor.

The National Research Council was also disgruntled with Pauling and wrote to criticize him for not living up to his promise to go to Berkeley. They had agreed to a delay but expected him to go there eventually. Pauling, however, was so excited by the prospect of the European trip that he put these problems aside.

The other complication was his new family. Ava Helen had fulfilled her ambition to become a mother, and Linus Pauling, Jr., was eight months old at the time. Ava Helen's mother, Nora Miller, argued strongly that they should leave the baby with her instead of taking him to Europe. It would be too expensive, as Pauling had only a stipend designed to support a single student. Also, the baby would be exposed to strange foods, germs, and possible seasickness. Nora Miller was eager for her daughter to have the chance to go to Europe and proud of her son-in-law's Guggenheim Fellowship. Ava Helen concurred with the decision to leave the baby with her mother, and Linus went along with it. He found it natural to defer to Ava Helen on decisions about child rearing.

Belle Pauling's health was deteriorating rapidly, but her children and their spouses were all accustomed to her chronic bouts with pernicious anemia, from which she had always recovered in the past. In any event, Linus's younger sister Lucile was living with her. But, while Linus and Ava Helen were in Europe, Belle did become terminally ill, and Linus's Aunt Goldie left her home and family in Texas to help Lucile take care of her. Finally the anemia caused a mental illness that made Mrs. Pauling too difficult even for the two of them to handle and she was sent to the state hospital in Salem. Grandmother Miller took Linus Jr. to visit her there, but the visits were a nightmare. The news of her death reached Linus and Ava Helen in Germany two weeks after it occurred. For years afterward, many of his acquaintances and relations in Condon spoke less about his two Nobel Prizes than about "the way he treated his mother." Linus and his mother had never been close and had become even more alienated by her efforts to prevent him from going to college.

For Linus and Ava Helen, these tug-of-war decisions established a precedent they followed throughout the remainder of their lives. Linus's work had priority over other obligations. His success was the primary goal of the family, and Ava Helen would make every effort to accompany him on any long trips, with or without the children.

Throughout this period, the new technique of X-ray diffraction was spreading like wildfire. Pauling was just one of many scientists attempting to determine the molecular structures of various elements using Bragg's technique. The structures of most elements and simple salts were discovered at this time, and scientists went on to attempt to determine the structure of more complex molecules. This was, however, enormously difficult—the "backward reasoning" problem, going from the diffraction pattern to the structure, was far more severe. In determining the structure of simple molecules, scientists were able to check out, in the laboratory, all of the structures that were possible, given the known characteristics of the elements that composed them. With more complex substances, which were of great interest to chemists, there were too many logical possibilities for scientists to check each. Finding the structure became largely a matter of guesswork. A scientist thought of a structure, then went to the laboratory to find out whether his model was correct. If not, he simply had to try another one. This was a lengthy and frustrating process, with a low rate of success.

Pauling was very good at this sort of work. He had a strong intuitive sense of the ways that atoms could fit together in molecules. Often, he worked with physical models, joining elaborate structures similar to a child's toys. Pauling's most important contribution, however, was not the particular structures that he discovered but his ability to codify and communicate the guidelines and assumptions he used in determining molecular structures.

In a paper published in 1928, "The Coordination Theory of the Structure of Ionic Crystals," Pauling proposed five "rules" that scientists could use in solving the structure of molecules in which the atoms were bonded because some of the atoms have a negative charge caused by gaining electrons from another atom they bond with while others have a positive charge caused by losing their electrons during the bonding process. Simple substances could be solved by standard methods, but more complex substances, such as say, mica ($KAl_2Si_{20}10(OH)_2$), had many plausible structures that differed only slightly in their nature and stability. Pauling's rules provided a simple algorithm for winnowing the good structures from the bad.

His five rules could not be proved to be correct on the basis of any definite evidence, but they were very useful. They were derived in part from Pauling's experience in identifying crystal structures and in part from his understanding of quantum mechanics and electrostatics. The use of these rules narrowed the number of possible solutions to the crystal structure of many complex molecules to a small number of choices, one of which could almost always be verified in the laboratory.

The first rule is that "a coordinated polyhedron of anions [negatively charged atoms] is formed about each cation [positively charged atom]," the distance between the anions and the cation determined by a certain formula. The second rule, the most important one, is the rule of "electrostatic valence": the state of maximum stability for an ionic crystal is the one that satisfies the following property: the valence of each anion, if one changes the sign, is equal to the sum of the strengths of the electrostatic bonds to it from the neighboring cations. Sir Lawrence Bragg, the co-originator of X-ray diffraction as a technique for structure determination, called this rule "Pauling's law" and considered it enormously important:

> The rule appears simple, but it is surprising what rigorous conditions it imposes on the geometrical configurations of a structure. In a silicate, for instance, each silicon atom is surrounded by four oxygen atoms. These atoms have half their valency satisfied by the silicon, and so are left with an electrostatic charge which is unity. . . . Aluminum within an octahedral group of six contributes one-half to each oxygen. Magnesium or ferrous iron within an octahedral group contributes one-third. Hence we may link a corner of a silicon tetrahedron to another silicon tetrahedron, to two [aluminum] tetrahedra, or three [magnesium] tetrahedra. . . . Proceeding to link tetrahedra and octahedra in this way, we find that very few alternative structures which obey Pauling's law remain open to a mineral of a given composition, and one of these alternatives always turns out to be the actual structure of the mineral. . . .
>
> This rule may be termed the cardinal principle of mineral chemistry.[4]

Pauling's rules, in many cases, predicted that certain structures could not exist, despite their apparent possibility according to the normal laws of valence, or bonding. Their empirical success was remarkable given their relative simplicity. This work was the first example of Pauling's unique knack for imposing order on apparent chaos. As he once said, "I like to take a very complicated subject where there is no order . . . and think about it for a long enough period that I can find some way of introducing order into it."[5] Time and time again, he conceived simple, intuitive rules that organized what, to everyone else, appeared to be a hopeless mess of complexity.

A great deal has been written about the philosophical oddities of the quantum world. Writers with a poetic bent see the uncertainty principle as a metaphor for the quandaries and imponderables of everyday life. This kind of metaphysical speculation never appealed to Pauling, however. For him, the challenge was to find the answers to very difficult research questions. He used quantum theory because it worked:

> I tend not to be interested in the more abstruse aspects of quantum mechanics. . . . I take a sort of Bridgmanian attitude toward them.

> Bridgman . . . would say that a question that does not have operational significance, that does not lead to an experiment of some sort or an observation isn't significant. I have never been bothered by the detailed or penetrating discussions about interpretation of quantum mechanics. In my (Messinger) lectures for example I discussed the matter of free will or determinism. There has been the contention that the uncertainty principle means that we are able to accept the concept of free will whereas, according to classical mechanics, free will was ruled out, in that the world was determined. If we knew the positions and moments, the velocities, of all the particles in the universe then we could predict the entire future history of the universe. My answer to this is that in fact the existence of the uncertainty principle does not affect the validity of this statement at all in any practical sense. Even if this were a classical world it would be impossible for us to determine the positions and moments of all the particles of the universe by experiment. But even if we did know them, how would we carry out the computations? We can't even discuss in detail a system involving, say, 10^{20} particles or 10^{10}, so I think it is meaningless to argue about determination versus free will, quite independent of the uncertainty principle.[6]

Regarding the debate among Bohr, Einstein, and Schroedinger about the philosophical significance of quantum mechanics, his attitude was dismissive. "I just couldn't get interested. The question in my mind was, 'Is quantum mechanics, wave mechanics, for example, sufficiently close to being correct so that if we solve the equation we'll get the right answers in relation to the properties of atoms and molecules?'—not even going beyond that to the nucleus, but just atoms and molecules."[7]

Pauling was impatient with philosophical speculations that focused on logical contradictions or paradoxes that supposedly needed to be resolved before science could proceed. He argued, "We have a tremendous amount of understanding of the universe and there are many striking phenomena that occur and that can be explained. The Reductionists say that as time goes by, more and more of the world will become explicable in terms of the parts that we do understand and I believe that theory."[8]

This pragmatic attitude may be disappointing to some people—after all, the classical view of the world had just been revolutionized! Absolute space and time were gone, and in place of them we had a shifting, nebulous world full of things that were both waves and particles, and were not quite there until someone looked at them. But Pauling's view was very much that of the physical chemist, concerned with the more focused problems of his own discipline. He was not concerned with vague speculations about the nature of the cosmos, nor did he care to be distracted by questions about the philosophical foundations of his thinking. He considered these issues largely a waste of time. For him the challenge was to make solid scientific discoveries about specific substances, specific kinds of matter. The fact that much of quantum theory seemed absurd from an everyday common-sense point of view did not disturb him in the least. He would accept any physical theory that accorded with the facts, no matter how odd it might seem to the philosophers, and adapt his intuition accordingly.

4

"An Extraordinarily Productive Scientific Kopf," 1926–1935

When Arnold Sommerfeld visited Caltech in 1924, he attended a small seminar in which Pauling presented models showing the structure of the water molecule, among other things. Sommerfeld had seemed impressed with Pauling's work, so Linus wrote to him at the University of Munich, and Sommerfeld responded by welcoming him to study at the university. Linus decided to make Munich his base in Europe.

Linus wanted to meet the famous European scientists, of course, but he was just as eager to meet the younger scientists who might be making the discoveries that would be famous in the future. A fellow graduate student at Caltech had mentioned that two Dutch scientists, Samuel Goudsmit and George Uhlenbeck, had discovered that the electron was spinning on its axis, not just circling the nucleus. This idea intrigued Linus, and he was eager to hear about their work firsthand.

The Paulings bid young Linus Jr. a tearful goodbye in Oregon, and went to New York, where they sailed on an Italian boat, the *Giulio.* At first Linus was seasick, but by the time they reached Madeira, on the way to Gibraltar and Algiers, he had adjusted and was able to enjoy traveling by sea. At a hotel at Naples, they saw a library across the street and Linus went to look for the most recent issue of the *American Chemical Society Journal,* in which the last paper of his thesis was to appear. His lengthy search ended in the office of a local professor, who seemed annoyed by his presence. He was too shy to mention that he was an author of one of the articles in the journal.

Linus and Ava Helen spent a month in Italy and stopped briefly in France, where Linus was too shy to look up Marie Curie. When they arrived in Munich, Sommerfeld didn't remember Pauling until he was reminded of the talk in Pasadena. Sommerfeld was gracious but aloof. They were invited to his home for tea and went on a few walks with him. Sommerfeld especially liked Ava Helen and, at his own expense, sent the second piano in his household to their boarding house in Munich.

Linus attended Sommerfeld's lectures on quantum physics, which met for two hours a week, and also Sommerfeld's two seminars. Pauling did not, however, find Europe more stimulating than Pasadena or Berkeley. The senior scientists were too busy to talk much with young foreign visitors, and Linus spent the bulk of his time working on his own, much as he would have in California. He discovered an error in a paper written by one of Sommerfeld's assistants and published a correction in *Zeitschrift fur Physik.* He then went on to write another paper, using the corrected method to develop the theory further.

Sommerfeld said that if Linus's second paper turned out as well as the first, he would like very much to submit it to the *Proceedings of the Royal Society.* Sommerfeld had only recently been made a member of the Royal Society, and only members could submit the work of others for publication. Sommerfeld also wrote to the Guggenheim Fellowship office, asking that Linus be given an additional six months in Europe, describing him as having "an extraordinarily productive scientific Kopf." Pauling accompanied Sommerfeld on a trip to a conference in Zurich, where he met Erwin Schroedinger, who had just that year published his own epochal paper on the wave mechanical approach to quantum physics. There was, however, little discussion of these ideas at the meetings and Linus spent most of his time working in the hotel room on his own papers.

While it was exciting to meet the world's leading scientists, Linus actually gained more from his contacts with scientists of his own age. Especially valuable was his meeting with two young physicists, Fritz London and Walter Heitler, who were working on the structure of the hydrogen molecule. Their classic paper, published in *Zeitschrift fur Physik* on June 30, 1927, was the first to show that the principles of quantum mechanics could be used to explain the bonds between atoms. It was, however, limited to the simplest of all bonds, the bond between two hydrogen atoms. Pauling's classic work on the chemical bond extended their ideas to deal with more complex molecules.

A side trip to Göttingen enabled them to meet Max Born and Werner Heisenberg as well as a young man whom they would later encounter in other circumstances, J. Robert Oppenheimer of New York City. Linus had never given up hope of going to Copenhagen to see Niels Bohr, although Bohr had not responded to his letters. Another traveling student told them that it was not necessary to have written permission to visit Bohr's institute in Copenhagen. One just went, and Bohr was receptive to all foreign visitors.

So the Paulings went to Copenhagen and found that Bohr was indeed courteous. The most valuable outcome of their visit, however, was a meeting with Samuel Goudsmit, the young Dutch scientist whose work Linus had heard of in Pasadena. Goudsmit told Linus about the manner in which doctoral examinations were given in the Netherlands. Instead of the boring quizzing of a student whose every item of knowledge was already well known to his questioners, the Dutch professors required each candidate to present twelve propositions and to defend them. Sometimes the candidate was permitted a thirteenth proposition, one that was bizarre or amusing.

Goudsmit's thirteenth proposition was on the Egyptian scarab, a black-winged dung beetle that was often represented in ancient Egyptian art and jewelry. He had discovered a new class of scarabs while working on his doctorate in Leiden and had learned to read hieroglyphics in order to do so.

If Goudsmit could learn Egyptian hieroglyphics, Pauling could learn to read Dutch, and promptly did so, at least well enough to translate Goudsmit's doctoral dissertation on line spectra into English and collaborate in expanding it into a book. As the young men worked and planned, their neglected wives entertained each other and once had tea with Mrs. Niels Bohr. They went to see the shops and museums and explored.

While the Paulings had enjoyed their European trip, Linus felt that it was easier to talk at length with noted European scientists in Pasadena than in Europe. The scientist on tour abroad, freed of daily routine, family obligations, and administrative duties and invigorated by the attention he received, was often more responsive. Linus and Ava Helen were ready to go back to the United States.

They returned to an exuberant two-and-a-half-year-old son who was strange yet familiar to their delighted eyes. His grandmother had prepared him well for their coming. He went running to meet them as they approached the house, shouting, "Mama!" "Papa!" The joy of the reunion was dampened only by the absence of Linus's mother and the coolness of the extended Pauling family, who still considered Linus's absence during her terminal illness inexcusable.

While Belle had died without ever realizing her ambition of seeing her son established in a regular job, Linus did have a job waiting for him, an appointment as assistant professor of theoretical chemistry and mathematical physics at Caltech. When he arrived back on campus, however, he found that "mathematical physics" had been dropped from his title. A few years later, the physics department objected to his teaching a course that used a physics textbook: *The Structure of Line Spectra* by Linus Pauling and Samuel Goudsmit. Pauling later speculated that the decision to drop the reference to physics in his title may have been influenced behind the scenes by the chemistry department, which did not wish him to wander too far into physics.

Despite these minor annoyances, however, Linus was generally happy at Caltech and remained there despite tempting offers from other institutions. In 1929, James Bryant Conant invited him to visit Harvard University for a week and offered him an associate professorship. But Linus did not like the atmosphere at Harvard—the graduate students were still working under conditions that he had found unacceptable when he was applying to graduate school, and he didn't feel he could encourage young students to go there. There were problems in the allocation of research funds, and professors of chemistry bought their supplies in a downtown store rather than pay the high prices in their own stockroom. Each professor had his own small group of students close to him, and there was little friendly cooperation between cliques. Not even the offer of a promotion enticed him to go to Harvard.

The University of California at Berkeley remained very attractive, however, and for five years, starting in spring 1929, Pauling lectured in both physics and chemistry at Berkeley for about two months each spring quarter. He had "wonderful arguments" with Gilbert N. Lewis. The rest of the year he returned to Pasadena. Another Pasadena-Berkeley commuter was J. Robert Oppenheimer, who, however, spent most of his time in Berkeley.

In 1930, the Paulings made a second trip to Europe to stay for six months, this time taking Linus Jr. with them. Linus had been working hard on the structure of mica and other silicate minerals and had hit snags in his thinking. He thought he might get help at the W. L. Bragg Laboratory in Manchester. Bragg had authored the book Pauling had taken with him for the summer before he went to graduate school. He found no solution for his specific problems in Manchester, however, so he did not stay long, but while he was there, he learned how to operate a Bragg X-ray spectrometer.

The Paulings went from Manchester to Munich. An old friend, Peter Paul Ewald, found them satisfactory quarters in a farmhouse in a village called Holzhausen at Ammersee, which was thirty miles from Munich. Linus hunched doggedly over his research for six days, and on the seventh, which was usually a Wednesday, he went to Munich to use the library at the Sommerfeld Institute. When he had collected his library data, he went to his favorite beer hall, the Spatenbrau, in downtown Munich, to enjoy an omelet and a stein of beer before he took the train home.

The farmhouse in Holzhausen at Ammersee had no cooking facilities, so he and Ava Helen had to walk a mile and a half to the village for their meals. Linus Jr. had a small bicycle that he rode as his parents walked.

Both Ava Helen and Linus Jr. learned German during those months. They swam in the lake, hunted mushrooms in the forest, and went rowing. Sometimes they went across the lake to another village. Most of the time the couple walked and read, but they made occasional trips to Munich for lectures and museums and concerts and the theater. Ava Helen was pregnant again.

While they were visiting Munich, Linus turned surgeon and performed a minor operation on Ava Helen, the description of which has impressed, and horrified, each ophthalmologist to whom it has been related. Ava Helen awakened one morning with a severe pain in her eye. Linus turned up the eyelid. An eyelash had turned backward, grown through the skin of the upper lid in what was nearly a complete circle, and penetrated the skin so that a sharp point was protruding and scratching her eyeball.

No minor measures relieved the intense pain. Because it was Sunday, finding a physician was difficult. Linus left Ava Helen holding her eyelid away from the eye while he went to a drug store to buy a pair of tweezers. With the tweezers and a razor blade, he was soon able to dissect out the eyelash and relieve her of the pain. The lid healed quickly. The operation was a success.

Quite by chance, while in Germany, Linus made a profitable discovery. He visited Hermann Mark in his laboratory in Ludwigshafen. Mark showed him the electron diffraction apparatus with which he and an associate had determined the structure of gas molecules of carbon tetrachloride and benzene during the preceding year. Pauling was overwhelmed with the possibilities of the new technique, which would allow him to determine the structure of molecules without at the same time having to determine the sometimes very complicated ways in which the molecules are arranged relative to one another in a crystal. Pauling asked Mark whether he would object to Pauling's doing some similar diffraction work in Pasadena, and Mark said that he did not. In fact, he was not planning to continue the work much longer.

By far the most important work to come out of this period of Pauling's career—and, indeed, Pauling's career as a whole—was the "valence bond" theory. This theory was a tremendous stroke of scientific imagination, which Pauling accomplished by combining his vast knowledge of practical and theoretical chemistry with the new quantum-theoretic ideas that he had learned firsthand from the European innovators. Crystals, the subject of Pauling's early work, present a very simple, unusual case of the general phenomenon of chemical bonding. Pauling wanted to move beyond this special case and use quantum physics to understand chemical bonds, not merely in crystalline structure, but in molecules generally.

According to quantum theory, we have said, particles were to be viewed in terms of quantum wave functions, as complex wave-particle combinations. Speaking informally, an electron could now be conceived as a sort of stationary wave—not a wave that moves, like an ocean wave, but a wave that stays in one place, interacting with other waves by the laws of wave motion.

In Schroedinger's interpretation of quantum physics, the wave is related to probability in that the "height" of the wave squared at a given point indicates the relative probability that the electron will be found at the location. In mathematics these waves are called *eigenfunctions;* in physical chemistry the preferred term is *orbital* when referring to only one electron.

To understand Pauling's ideas about the chemical bond we must scrutinize these "waves" a little more closely. The waves that we are familiar with in everyday life are one- or two-dimensional. We might observe a one-dimensional wave, for example, when two people hold the ends of a rope and shake it. If the rope were frozen at one position, it would resemble the one-dimensional wave illustrated in figure 5. The crests and troughs of the wave are separated by a region of calm in between. In the drawings, we have made the positive (or up) part of the wave heavily shaded and the negative (or down) part of the wave lightly shaded.

Two-dimensional waves are like waves in an ocean, which spread out over the surface of the water. To portray these waves, the artist uses variations in the shading to give the drawing the appearance of depth, as in the second wave in figure 5. The peaks and troughs of the wave are high and low points on the water, while the zones of calm are smoother patches of water in between.

Electron waves, however, are three-dimensional. An example of a three-dimensional wave is a shock wave caused by an underwater explosion. It may also be stronger in some directions than in others, and it may vary in intensity, depending on how far it is from the center of the electron. These waves are difficult to draw on paper, because you really need to convey the variations in intensity as well as the three spatial dimensions. We have drawn cross sections of the waves and used shading to reflect differences in the positive or negative value of the wave at a given moment.

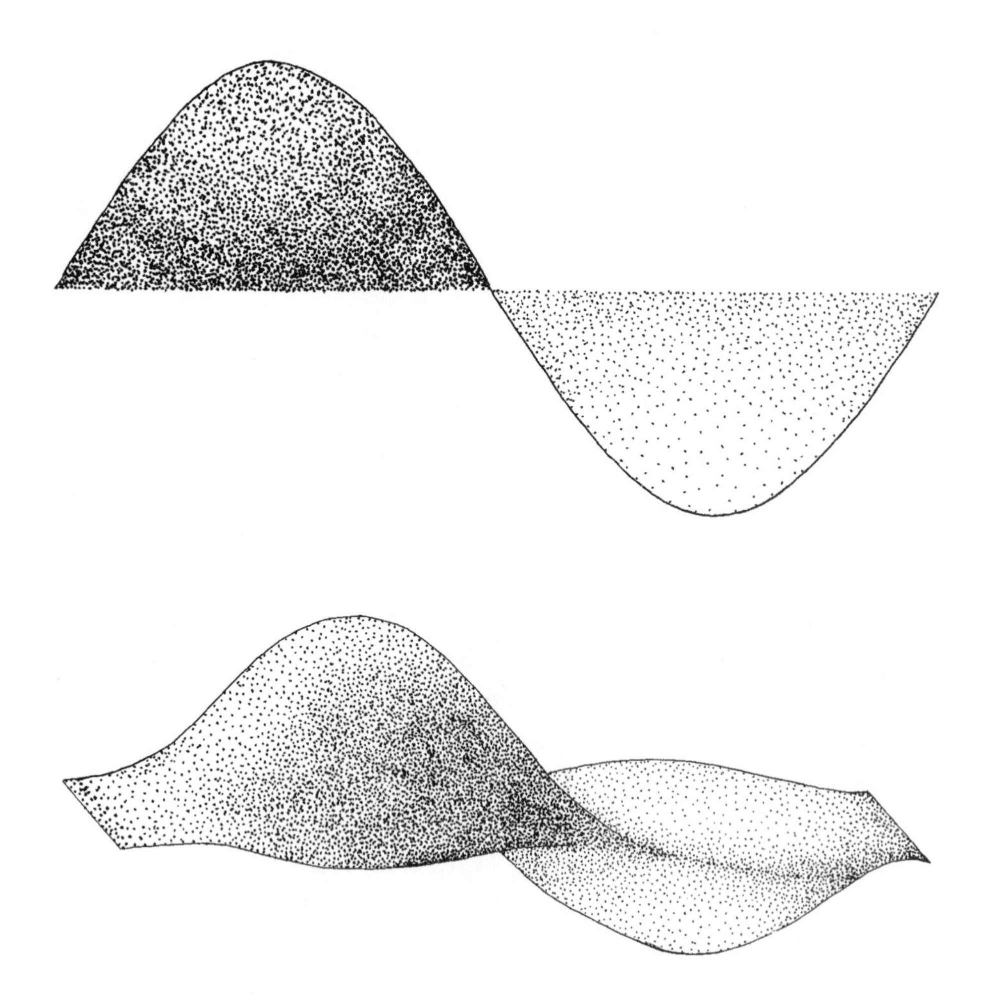

Figure 5. Waves in two and three dimensions.

The simplest kind of three-dimensional stationary electron wave is the spherical wave, which is the same in every direction. Figure 6 is a drawing of a cross section of the wave, but it is easy to visualize it in the three dimensions of a sphere. The shading represents the peaks and troughs in the positive or negative value of the wave.

The hydrogen atom contains only a single electron, the waves of which can be described as a sphere consisting of either a single crest or a single trough. A lithium atom, on the other hand, has three electrons, one farther out from the nucleus than the other two. The outermost electron, in this case, possesses both a crest and a trough. The outermost of the sodium atom's eleven electrons can be described as a spherical wave with one trough and two crests. The properties of an atom are largely determined by the waves that are present on its outermost shell. Atoms with spherical electron waves on the outside—such as hydrogen,

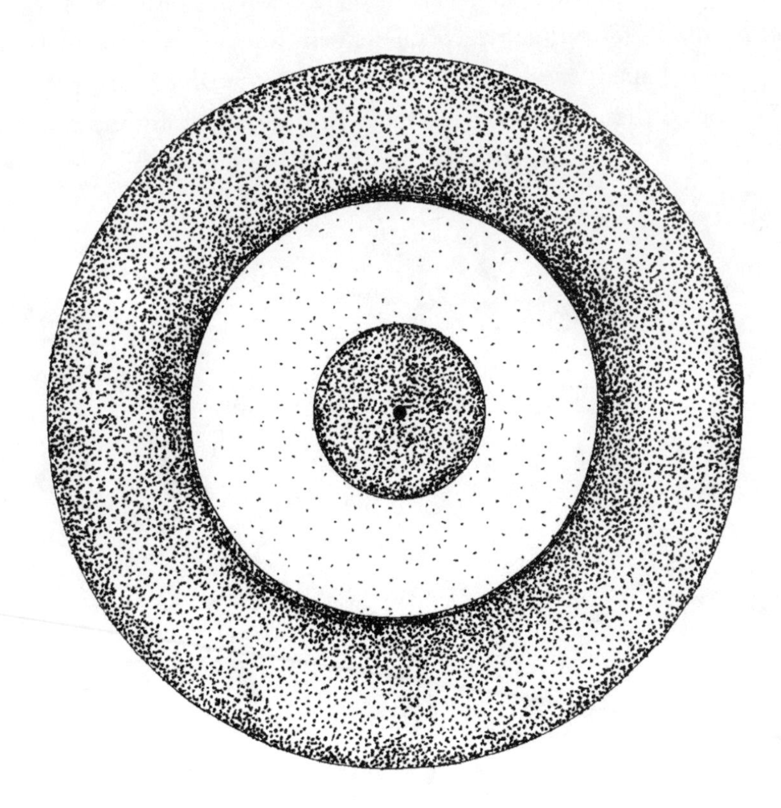

Figure 6. A spherical wave.

lithium, and sodium—have a fundamental symmetry: they can form bonds with other atoms in any direction equally well.

In technical language, a spherical electron wave is an *s orbital.* More interesting are the p orbitals and d orbitals, which, as Lionel Salem has observed in his eloquent popularization *Marvels of the Molecule,* are well described as "figure-eight waves" and "cloverleaf waves." As shown in figure 7, these give the electrons more space to move about and allow a greater variety of atomic structures. Carbon, nitrogen, and oxygen are based on figure-eight waves, and metals such as chromium and iron are based on cloverleaf waves.

One might think that this process could go on and on forever—why not figure-nine waves, maple leaf waves, creeping ivy waves? But science is rarely so straightforward. What happens instead is that, in order to create better bonds with other atoms, atoms can change the shape of their waves. This leads to what are called s-p hybrids, or, in Lionel Salem's visual terminology, "petal-shaped waves." The formation of these petal-shaped waves is the *hybridization of orbitals.* The paper that first reported these waves was arguably Pauling's most important contribution to science. Entitled "The Nature of the Chemical Bond: Application of Results Obtained from the Quantum Mechanics and from a

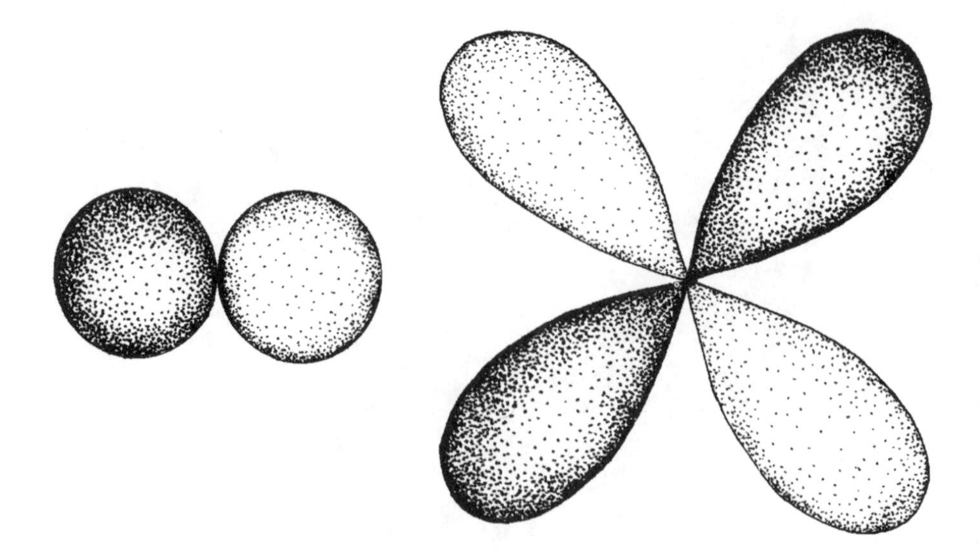

Figure 7. Figure-eight and cloverleaf waves, depicting s and p orbitals.

Theory of Paramagnetic Susceptibility to the Structure of Molecules" and appearing in the *Journal of the American Chemical Society* in 1931, it was the contribution of which Pauling himself was most proud.

Before describing the details of Pauling's theory, let us consider a simple example. In the late nineteenth century, the Dutch scientist J. H. van't Hoff and the French scientist J. LaBel had independently proposed that carbon forms four bonds in a tetrahedral shape. This was one of the first moves toward a fully three-dimensional view of molecular structure. To see the importance of this model, consider the two structural diagrams for dichloromethane, shown in figure 8. Viewed two-dimensionally, as they are in the diagrams on the top, these appear different and should have different chemical properties. But they actually reflect the same structure when viewed three-dimensionally, as they are in the two drawings on the bottom. This particular molecule has only one three-dimensional structure, that of a tetrahedron, but in a molecule where the four atoms linked to the carbon atom are all different, two different three-dimensional structures are possible. An example is the amino acid serine. Molecules like this are called *dissymmetric:* they have two different forms, which are by convention called "left" and "right." In the case of serine only the left version occurs in nature, but the right version can be produced in the laboratory. The study of such molecules is *stereochemistry.* One result of Pauling's discovery of the hybridization of orbitals was to place these various phenomena on a firm quantum-mechanical footing, thus bridging the gap between the chemist's understanding of molecular structure and the physicist's understanding of the inner workings of the atom.

A carbon atom, in the quantum view, has three figure-eight waves and one spherical wave. From these, by the hybridization of orbitals, it can produce four petal-shaped waves that point in four different directions. The petals, each consisting of a small trough and a large crest, mark out the corners of a four-faced pyramid or tetrahedron. This shape is not observed in the isolated carbon atom, but it becomes apparent when the carbon atom is surrounded by other atoms—say, by hydrogen atoms. Each of the four outer electrons of the carbon atom binds up with the single electron of a hydrogen atom, thus forming a stable tetrahedral structure that is called *methane.* In this way, according to

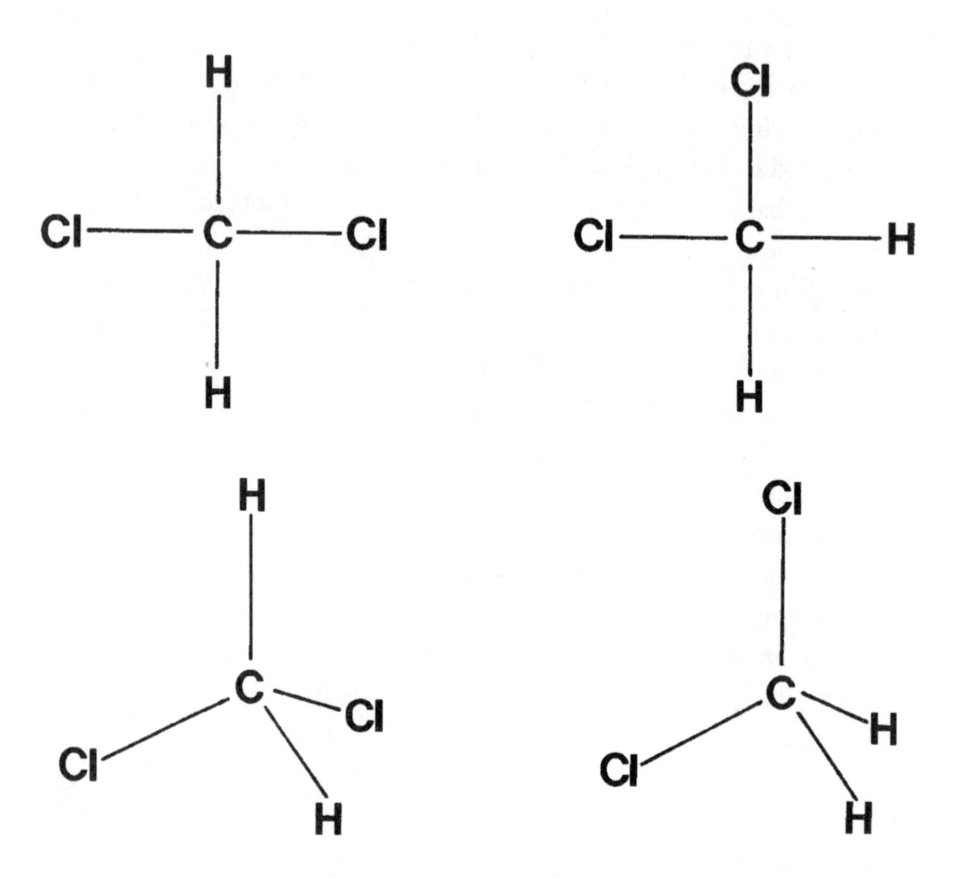

Figure 8. Different two-dimensional diagrams for dichloromethane, representing the same three-dimensional structure.

Pauling's ideas, orbitals influence one another to make molecular structure possible.

Pauling was building on a body of earlier worker regarding the use of quantum physics to study chemical structure, most notably a 1927 paper by Walter Heitler and Fritz London. Also, parts of his work were independently duplicated by J. C. Slater, thus leading to the name HLSP (Heitler-London-Slater-Pauling) theory, which Pauling used in his own writing to refer to the valence bond theory. While Heitler and London's contributions are frequently highlighted in European writing on the subject, Americans, particularly chemists, tend to associate the theory primarily with Pauling. But the truth is that, while Heitler and

London's work was important, Pauling was the first to devise a systematic method for applying quantum-mechanical concepts to complex molecules. It was Pauling's ideas that made the crucial link between the atomic and molecular realms.

It is perhaps difficult for the nonscientist to appreciate the magnitude of this achievement. Even a molecule as "simple" as methane, considered as a system of elementary particles, was far too complex to be analyzed mathematically using the equations of quantum physics. One might say that deriving the behavior of a molecule by quantum physics is like deriving the behavior of a group of people from a knowledge of the personalities of the individual people. In both cases, certain rough predictions can be made easily, but gaining detailed understanding is very difficult, because many subtle interactions are at play.

Today one can obtain rather good results about molecular structure from computer simulations, but in the 1920s and 1930s computers did not exist, and one had to rely entirely on human ingenuity and mathematical tables. Pauling's theory of the chemical bond consisted of six rules, three of which followed fairly directly from the mathematics of quantum theory as applied to hydrogen, helium, and lithium atoms, and three of which were pure inspiration. Each of these rules was stated in mathematical form. It is possible, however, to express the meaning of the rules in ordinary language, although much of the precision is lost.

The first three rules, roughly speaking, are as follows: First, electron-pair bonds are formed by the interaction of two unpaired electrons, one on each of the two bonding atoms. Second, the spins of the electrons must be opposed when the bonds are formed, so that they do not contribute to the magnetic properties of the substance. And third, the two electrons that form a shared pair cannot take part in forming additional pairs. These rules systematized the understanding of chemical bonding that was emerging from the rapidly developing quantum theory.

The next three rules were fundamentally novel; they may represent Pauling's greatest stroke of genius. They exemplify, more than any other single discovery, the extraordinary chemical intuition that, in one area after another, led Pauling to simple and elegant explanations of extremely complex phenomena. The rules were justified, in the 1931 paper, by sketchy mathematical and qualitative arguments; their real justification lies in the numerous chemical structures that have been correctly inferred from them.

The fourth rule states that the most important terms in the equations for the electron-pair bond are those involving only one quantum wave function from each atom. This is a mathematical approximation of the type often made by physicists: one ignores interactions that are "small" in magnitude in order to derive tractable equations. The trick is always to make the right approximation—not to overlook the important points. Pauling's fourth rule was inspired by the nineteenth-century idea of valence; it was "right" in the sense that the interactions that it ignored were in many cases insignificant.

Pauling's fifth rule was the greatest innovation: it states that, generally speaking, stronger bonds are formed by orbitals that overlap more with orbitals of the other atom. So, if there are two orbitals competing for a bond with a certain atom, the winner will be the one that overlaps more with that atom. In addition, the direction of the bond formed by an orbital will tend to be the same as the direction that the orbital is concentrated in. There was nothing in the old idea of valence to suggest this fifth rule, because the idea of "overlap" was a new one, a corollary of the idea of a quantum wave function. However, despite its lack of precedents, the rule smacks of common sense. Greater overlap makes a stronger bond, and the bond is in the direction of the orbital that is bonding—these are very natural, intuitive conclusions.

Finally, the sixth rule states that, between two orbitals concentrated in approximately the same direction, the stronger bonds will be formed by the orbital closer to the nucleus of the atom, which corresponds to a lower energy level for the atom. This rule is mathematically similar to the fifth rule but tends to have fewer direct applications.

The fifth rule implies the petal-wave picture illustrated in figure 9. The petal wave is formed by the "stretching" of the orbitals of one atom in the direction of another, and this is exactly what is guaranteed by Pauling's fifth rule, which, out of the space of all possible orbitals, rewards those orbitals that overlap with the other atom most. This is the *hybridization of orbitals*—the ability of bonding, in itself, to affect the form taken by the orbitals of an atom. If one wished to wax poetic, one might say that atoms reach out to each other, distorting the quantum wave functions of their electrons in precisely the most effective way to "grab" each other. In this way atoms join together to make molecules, the basic elements of matter. This is a natural, intuitive idea, almost visceral in its simplicity, and it cuts through the complexity of interacting quantum wave functions in a most remarkable way.

This work culminated in a series of important papers, beginning with "The Nature of the Chemical Bond" in 1931, and is described in detail in Pauling's book *The Nature of the Chemical Bond.* The pivotal concept in the theory is the highly technical, highly innovative idea of the hybridization of orbitals, based on the concept of resonance among different electrons.

The basic idea for this paper came to Pauling in a flash of insight, after many days of struggling with complex mathematical details models:

> Finally, in December, 1930, one day I thought of a way to get around the mathematical difficulties. A simplification which made it very easy to get the results. And I was so excited and happy, I think I stayed up all night making, writing out, solving the equations which were so simple that I could solve them in a few minutes. Solve one equation and get the

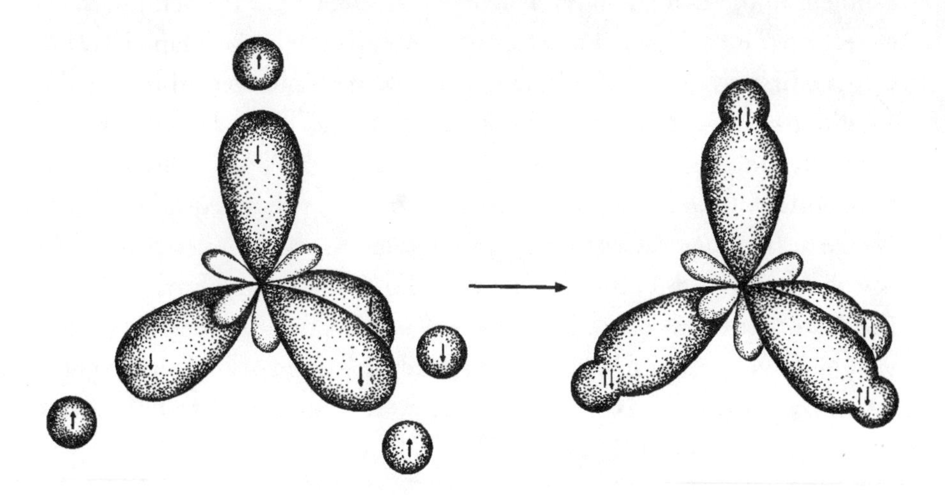

Figure 9. Petal waves, depicting the hybridization of orbitals in the carbon atom, as it bonds with four hydrogen atoms to form methane. This is an exemplary illustration of Pauling's theory of the chemical bond.

> answer and solve another equation. . . . I just kept getting more and more euphorious as time went by, and it didn't take me long to write a long paper about the nature of the chemical bond. That was a great experience.[1]

The 1931 paper combined chemistry and physics to an unprecedented extent. And, years later, Pauling remembered believing that it would have the journal editor, Arthur Beckett Lamb,

> buffaloed. . . . [He] thought, "What referee shall I send this paper to? It has to be somebody who has a good knowledge of chemistry . . . but also has a thorough understanding of quantum mechanics, and I can't think of anybody of that sort," anybody who might be said to be my peer. He [thought], "Well, past experience has shown that this author knows what he's writing about, so I'll just go ahead and publish the paper."[2]

The paper appeared just seven weeks after it was submitted.

The original 1931 article was followed within the next three years by many other articles refining and developing Pauling's model of the chemical bond and applying it to numerous other substances.

Pauling's achievement was quickly recognized by the scientific establishment in the United States. Late in the spring of 1931, he was selected to receive the Langmuir Prize from the American Chemical Society. He was the first recipient of this award, which was intended to honor the most promising young research chemist in the United States. Overnight, Pauling became a celebrity. His office and his home were invaded by reporters. Caltech and Pasadena were proud of the accomplishments of one of their favorite sons. The wire services spread the news throughout the country and abroad. The *New York Times* and *Herald Tribune,* the *Christian Science Monitor, The Nation,* and the *Portland Oregonian* were among the publications that spoke of Pauling as "the rising star who may yet win the Nobel Prize." They were quoting the president of the American Chemical Society.

The *New York Times* told how Albert Einstein, while visiting the Pasadena campus in 1931, asked many questions of Pauling at a seminar, confessed his lack of understanding of the chemical bond, and apologized for taking so much of the speaker's time. The *Portland Oregonian* speculated that if only ten men in the world could understand Einstein's theory of relativity, there must be even fewer who could understand Pauling's work. This was not actually true of Pauling's or Einstein's work, of course. Pauling's work was readily understood by other specialists in the field of molecular chemistry. In fact, unlike many other leading scientists, Pauling had a great gift for making his ideas intelligible and did so frequently in lectures.

This fundamental research culminated with the publication of his classic work, *The Nature of the Chemical Bond and the Structure of Molecules and Crystals* in 1939. This book, with revisions in 1940 and 1960, became one of the mainstays of modern chemistry. In 1947, Pauling's textbook *General Chemistry* was published; it became one of the best-selling college texts. Revised editions of this book, and the related *College Chemistry* for liberal arts students, continued to appear throughout the 1950s and 1960s. All these later works were based on his pioneering research in the 1920s and 1930s.

What is the status of the Heitler-London-Slater-Pauling valence bond theory today? The important point to remember is that, intuitive as they are, Pauling's rules are only approximations. All this work, all these deep intuitions and wonderful leaps of the imagination, simply approximate the solutions to certain equations, which are too difficult to solve exactly! Today Pauling's "valence bond theory" approximations are not often used; they have been largely supplanted by a different approximation scheme, the *molecular orbital theory,* developed by Robert Sanderson Mulliken, which can be solved more easily by computers. But both of these theories are just approximations to the real solutions of the equations of quantum mechanics, which remain intractable except for simple cases.

A simple example will illustrate the difference between the two points of view. The molecule in question is benzene, which consists of six carbon atoms linked in a hexagon, with hydrogen atoms surrounding them (figure 10). This molecule was a bit of a paradox in early organic chemistry because it appeared to be symmetric, but there was no apparent way of representing the internal bonds in a symmetric way. This is because the carbon atom uses only three of its four bonds in establishing the basic structure shown in the diagram. The fourth bond is used to establish a double bond with one of the other carbon atoms. This can be done in two different ways, as shown in the second part of figure 10. The German chemist Friedrich Kekulé came up with a solution to the puzzle: he proposed that the molecule oscillates between the two different forms. The valence bond theory goes this one better and, in good quantum form, assumes that the molecule is a hybrid, or superposition, of the two possible forms, each represented by one wave function. The total wave function is an average of the two. This phenomenon is also known as *resonance:* not an average in Kukelé's sense but a more stable structure in which the mobile electrons interact with more than two nuclei.

The molecular-orbital theory, on the other hand, offers a different solution to the problem. Rather than focusing on the bonds between the atoms, it starts with the electrons surrounding the atoms themselves, as shown in part (a) of figure 11. It postulates a number of different bonding orbitals, of which part (b) shows the simplest. The two doughnut shapes represent regions of electron density; the key point is that the six

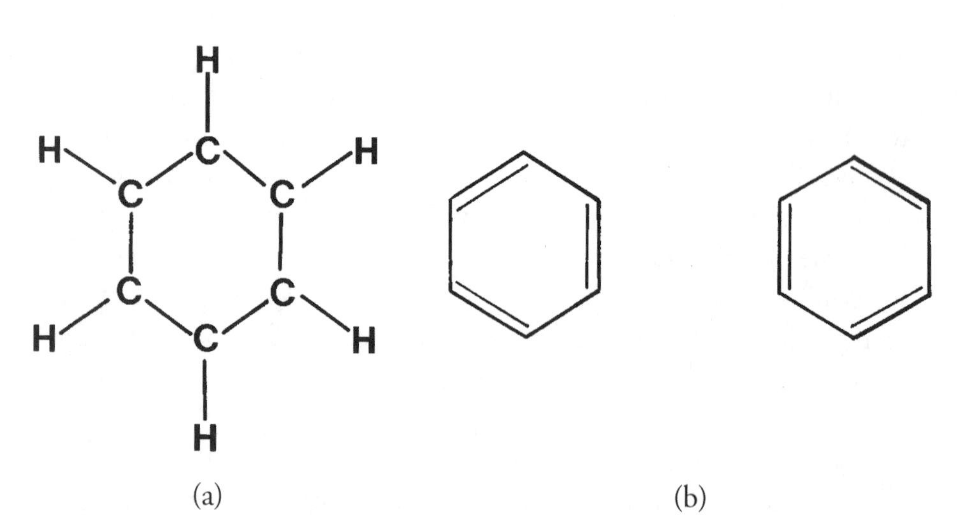

Figure 10. (a) The structure of the benzene molecule, which requires only three of the four bonds possessed by each carbon atom. (b) The two possible configurations of the benzene molecule; each double line represents a double carbon-hydrogen bond. Neither configuration in itself is consistent with the empirical properties of benzene. According to Pauling's theory, the actual molecule is a blend or "superposition" of the two configurations.

electrons in these regions of density are not associated with the bonds between any particular atoms. Instead, they range across the whole ring. The intuitive advantage of the valence bond theory for chemists is obvious. It corresponds to the way they are used to thinking about molecules.

Although Robert Mulliken and Linus Pauling were adherents of com-

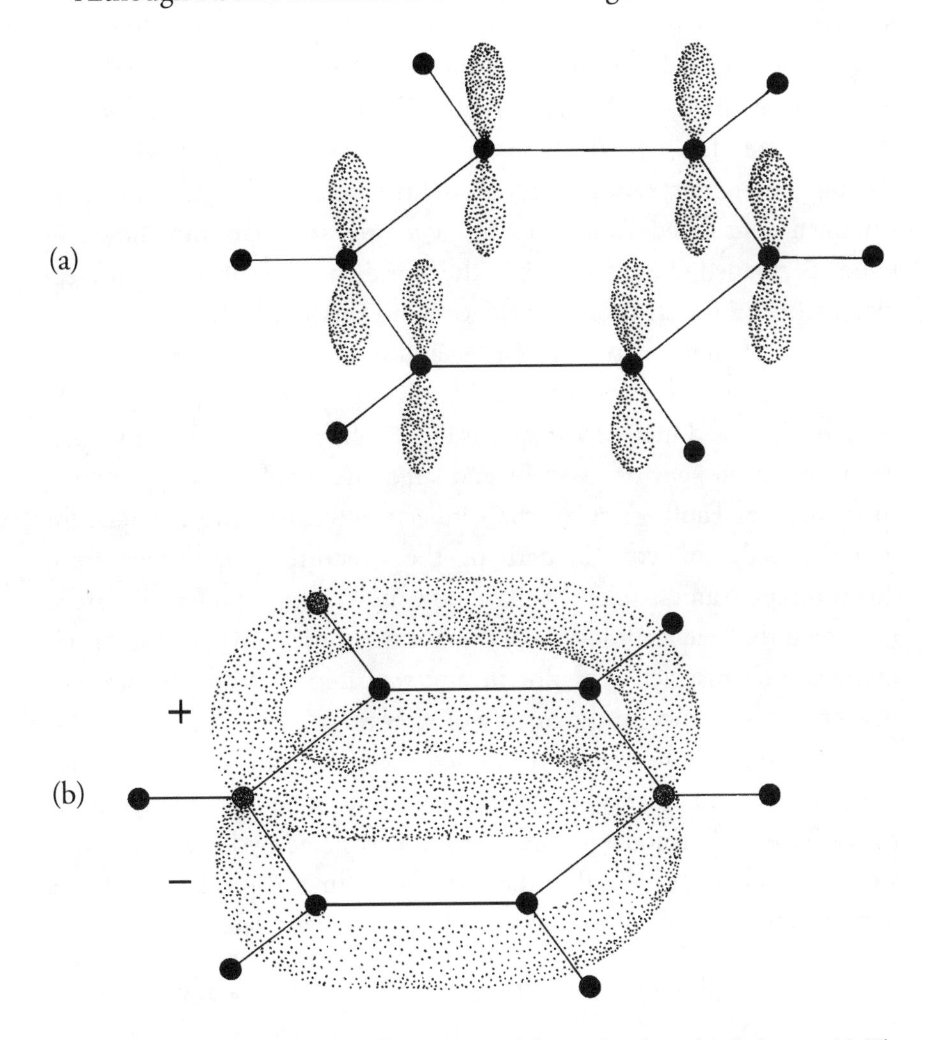

Figure 11. The benzene molecule as depicted by molecular-orbital theory. (a) The unused p orbitals on the carbon atoms are represented by excessively narrow lobes. The overlap between these orbitals leads to three bonding molecular orbitals of type i. (b) The simplest bonding molecular orbital of type i. Note the difference from the valence-bond theory explanation of the benzene molecule: here an orbital surrounds the whole molecule, instead of a superposition of different bonds between particular atoms.

peting theories for many years, they and their wives were quite friendly on a personal level. In his autobiography, Mulliken remembers that when he visited Pasadena with his wife and year-old daughter in 1935, "Linus and Ava Helen one day took us for a memorable expedition into the desert." When Mulliken was elected to the National Academy of Sciences, he "was pleased that Ava Helen Pauling greeted me affectionately, even though Linus, who was already a member, and I were rivals in our ideas about molecular structure." In 1948, Mulliken and the Paulings attended a party during a conference in Paris, at which the Paulings' daughter Linda put on "a little solo dancing exhibition." Mulliken considered Pauling to be "a master salesman and showman [who] persuaded chemists all over the world to think of typical molecular structures in terms of the valence bond method."[3]

Because of the fundamental intractability of the quantum-mechanical equations for large molecules, there is no rigorous way to calculate the error involved in approximations like Pauling's valence bond theory, except in simple specific cases. By and large, the proof is in the pudding. In the case of Pauling's rules, the verdict is clear: they were crucial for several decades of work in deriving the structure of molecules from quantum mechanics. Scientists have now mostly moved on to situations for which the valence bond theory is not the most convenient approximation, but this fact does not in any way detract from the theory's immense importance in the history of science.

Pauling himself, however, remained tenaciously committed to valence bond theory, and especially to the concept of resonance underlying the hybridization of orbitals. According to Robert Paradowski, a historian of science who has specialized in the work of Pauling, this may have had a negative effect on his productivity in his middle and later years:

> When [Pauling] is convinced of the value of a scientific idea he clings to it tenaciously and uses it boldly. Resonance was just such an idea. Early in his career he became convinced of its power and efficacy. Continued success in applying this idea to a great variety of chemical problems confirmed him in this attitude. Because of the strength of the resonance concept in his thought, he was correspondingly less aware of, or at least attracted to, the growth of new bonding ideas. Confronted with new problems, he tended to reshape them in terms that he understood and approved.[4]

But if at times Pauling clung too strongly to the valence bond theory, this is perfectly understandable, for the theory represented his own intuitive feeling regarding the nature of the molecule. The theory, through its many applications, served to pass Pauling's profound intuition on to others. Pauling's six rules were a first demonstration of the viability of quantum mechanics as a foundation for understanding the molecular structure of matter, and because of this they are a landmark in the history of science. He squeezed as much rigorous information as could be obtained out of the quantum theory and supplemented this information with simple, imaginative ideas that led to tremendous practical results. Today, as scientists struggle to understand the behavior of various complex systems, Pauling's theory of the chemical bond stands out as a paradigm case for the study of complexity.

Surveying Pauling's life and personality during these years of tremendous discovery, two features are immediately apparent: first, a surpassing devotion to his scientific work; and second, an unusual gift for explaining that work to his peers, for "spreading the word."

Pauling's flair for exposition accounted for a fair part of the influence of *The Nature of the Chemical Bond.* His crystal-clear language made it easy to overlook the occasional gaps in the mathematical explanations. It also served him well on many other occasions, for example, at the American Chemical Society meetings that were held in Buffalo in September 1931. Most of the two thousand participants turned out to hear the Langmuir Prize ceremonies and Pauling's lecture. All went well until Pauling called for his slides to be shown. The fuses blew and the hall was plunged into darkness. The hall was pitch black; all Pauling could see was the glow of cigarettes burning throughout the hall. Without perceptible pause, he went on speaking. He didn't need his slides or his notes to expound on the nature of the chemical bond. The speech was forceful, coherent, full of rich imagery and even humor. For many years, old-timers in chemistry remembered that night as a unique experience in their lives as well as in his.

Pauling was made a full professor in 1931, soon after he received the Langmuir Prize. He was an outstanding, brilliant lecturer in the classroom, even in freshman introductory classes. Frequently dressed in tropical sports shirts obtained on trips to Hawaii in the 1950s and 1960s, he drew the students into the material by sharing his thinking processes with them. He would often walk into the lecture room, perch himself on the laboratory table, and begin, "As I came here, I was thinking of how Professor (so and so) and I differ about the structure of (such and such)." He would then draw diagrams, present models, define terms, draw analogies, all to make his ideas clear. Pauling would then turn to the theory of the other scientist and make that position equally clear.

Pauling and his assistant spent hours making precise three-dimensional models of primary crystal structures from tennis balls and geometric forms that they pasted on wire frames to represent molecules. When he lectured in the classroom about the bright blue benitoite, a crystal rarer than diamond, he displayed a huge model of a molecule of benitoite, as he had determined it in the laboratory. Students were suddenly behind the scenes with a leading actor in a drama of discovery. The world of the interior of the molecule displaced science fiction as an exciting mystery for many a Caltech undergraduate.

But, despite his unusual flair for exposition, teaching was never more than an avocation for Pauling. Research was the only activity that commanded the full use of his talents, keeping him at his desk or in the laboratory long after midnight or during sunny weekends when other Pasadena citizens were enjoying the beach or the desert in bloom. His work was not compulsive or burdensome, but engrossing and enjoyable. The observer often felt that Pauling was "messing around" with his papers and equipment rather than working seriously. When a task became tiresome, he put it aside and worked on something else. He worked to get results, not to demonstrate his willpower, and felt no compulsion to finish a task for the sake of finishing it. Often, the solution to an obstinate problem would come to him later at a time when he was busy with something else or just relaxing.

This work style was tremendously productive. Pauling published his one-hundredth scientific paper, a "Note on the Interpretation of the Infra-Red Absorption of Organic Compounds Containing Hydroxyl and Amino Groups," in the *Journal of the American Chemical Society* in 1936. His first paper, "The Crystal Structure of Molybdenite," had appeared in the same journal in 1923. Throughout the 1920s and

1930s, he and his coauthors published numerous articles describing the structures of crystals such as hematite, corundum, barite, cesium triiodide, potassium cyanate, brookite, potassium chloroplatinate, chalcopyrite, ammonium hydrogen fluoride, calcium boride, Swedenborgite, cesium aurous auric chloride, enargate, as well as articles about the structures of groups of crystals. In all, 225 molecules were described in papers published by Pauling's laboratories—a remarkable record.

When Arthur Noyes died in 1937, Linus Pauling lost a father for the second time. The chemistry department was a lonely place without Noyes, and Linus was happy to accept an offer to visit Cornell University for a year. Each year, Cornell invites to its campus an outstanding scientist as George Fisher Baker lecturer. Linus and his family left for the Ithaca campus in fall 1937. There he gave a series of lectures to which the general public, faculty, and students were invited. A group of assistants gathered to work under his direction on studies of the magnetic properties of substances and their relation to chemical bonding.

When Pauling returned to the Caltech campus on May 25, 1937, he went as Noyes's chosen successor. He was appointed chairman of the chemistry division and director of the Gates and Crellin Chemical Laboratories. The California Institute of Technology was experiencing a period of rapid expansion during these years, and Pauling had no trouble finding the resources needed to build an outstanding chemistry department. His old acquaintance and fellow chemist, James Bryant Conant, who had once been at Caltech and was now president of Harvard University, returned to Pasadena to lecture and to see what was going on in the Crellin Laboratory. He rather envied Pauling. Now that Conant was an administrator he had no time for research, whereas Pauling still managed to find time for his own work. In Conant's younger days, he had been very close to arriving at a theory of the chemical bond himself but had stopped short of the final insight. He liked what he saw at Caltech. "You have the leading chemistry department in the USA," he told Linus.

Pauling was an impressive figure on campus. He seemed to enjoy both his status as a world famous scientist and his power as a laboratory director. Ava Helen saw to it that he was dressed in stylish business suits, and he was quite formal in his language and demeanor. The graduate students held him in awe. Oregon State University chemistry professor Kenneth Hedberg, who was a graduate student at Caltech in the late 1940s, remembers that one day Pauling noticed a key chain on his desk attached to a small viewing device. Pauling picked it up and looked through it, and Hedberg and the other students wondered how he would react to the picture of a beautiful woman, completely naked, standing on a large black rock in the middle of a rushing mountain stream. Playing the scene for all it was worth, Pauling peered intently for a few seconds, put the device down, and said, "Basalt!" as he walked out of the room. No one else had noticed the rock the woman was standing on.

The expanded facilities at Caltech made it possible to buy the latest equipment and to attract gifted professors and graduate students. Pauling borrowed a water-cooled magnet from George Ellery Hale's private laboratory so that E. Bright Wilson, Jr., could do some experimental work in magnetochemistry, the study of the magnetic properties of chemical substances. Hugo Thorell, who later became a Nobel Laureate, traveled from Sweden to work on myoglobin, hemocyanin, and cytochrome c. Robert Corey, who had gone to Caltech because Pauling was there, worked closely with him and in 1938 published a paper that reported the first detailed structural determination of a peptide. In 1936, Pauling and Alfred Mirsky published an article on the structure and properties of native and denatured proteins. In 1939, Gustav Albrecht and Corey reported the first structural determination of an amino acid. In 1940 Pauling published a theory of the molecular structure of antibodies and the nature of serological reactions. Hundreds of papers, many of which reported major "firsts," flowed from the Caltech chemistry laboratories. Pauling had proved himself, and he was living every scientist's dream.

5

Proteins, Politics, and Passports, 1935–1954

Linus Pauling entered middle age with an enviable record of accomplishment. He was a successful administrator and a popular lecturer and could have continued in a rewarding career as a scientist. Such a life might not have matched the challenges of his youth, when he was struggling to achieve a goal that was not shared by his mother or other relatives, but it would have given him more time to enjoy his own family. As the middle of Pauling's century approached, however, he made two major changes in his life. He became passionately involved in public affairs, and he shifted the main focus of his scientific work.

By making these changes, Pauling stayed at the cutting edge of scientific research, where he had the chance to make historic discoveries. He also allowed himself the time to act on his strong feelings about social issues, feelings he had not had time to develop as a young scientist. He was quite effective in the social policy arena, most impressively by forcefully standing up to McCarthyism at a time when many others were taking cover. The energy he put into these political pursuits, however, inevitably interfered with his concentration on science.

Just as, over the 1920s and 1930s, Pauling had gradually moved from the study of crystals to the study of more complex chemical bonds, in the mid-1930s he began moving from physical chemistry proper to the newer and more exciting area of chemical biology. In 1934, he became interested in hemoglobin and applied to the Rockefeller Foundation for a three-year grant to study it. The foundation had made it clear that it would be more enthusiastic about supporting his work if it dealt with biological topics that might have medical applications. In 1938, the foundation awarded Pauling and Caltech a large grant to support their research in the area of biochemistry.

X-ray diffraction techniques had advanced and now allowed detailed study of large organic molecules such as proteins. Pauling turned his formidable abilities to the problem of uncovering the molecular structures underlying biological systems. In doing so he, more than any other single scientist, helped to lay the foundations for modern molecular biology. In some areas he saw relationships that were not picked up on by other researchers until the mid-1980s.

It should not be supposed, however, that Pauling focused entirely on chemical biology during this period. While the average scientist works in one specialty for his entire career, and the exceptional scientist may spread himself among two or three different topics, Pauling's publications span the field of physical chemistry. Throughout Pauling's life, whatever the primary focus of his current research, he published occasional papers on related scientific topics that happened to catch his interest. Examples are his theory of anesthesia and memory, discussed in chapter 7, and the series of articles on the structure of atomic nuclei he published in the 1960s. This was a consequence of his "messing around" approach to research—he rarely hesitated to follow his intellectual whims.

Despite Pauling's eclectic approach, his primary research reflected a single-minded focus on fundamental questions. His work on molecular structure provided a systematic foundation for the study of crystals and placed mineral chemistry on a whole new footing. His theory of the chemical bond revolutionized the relationship between quantum physics and molecular chemistry. And in Pauling's later work on aspects of biochemistry, such as his research on hemoglobin, he was pioneering a new scientific specialization, the concept of molecular disease. Such specialized research was often in the service of some grand conceptual scheme.

Even in his primary research areas, Pauling was invariably only part of the story; he had many collaborators and competitors. He mentored a number of brilliant graduate students and postdoctoral fellows, who often made major contributions to the key discoveries that became associated primarily with his name. He never invented a theory "all his own," on the order of, say Einstein's relativity theory, or Dalton's theory of chemical structure, or Schroedinger's or Heisenberg's equations for quantum theory. Even when he moved into new areas, such as molecular biology, Pauling was never a true loner; he was always following trends.

But in this respect Pauling is in very distinguished company, for science in the twentieth century is not the solitary pursuit that it was in earlier times: one of the most significant trends in modern science is the emergence of large-scale laboratory research. Pauling was very much at home in laboratory work, able to grasp the implications of other scientists' findings, to collaborate with colleagues and graduate students, and to move rapidly from abstract mathematics to concrete experiment to wild speculation. Though some of Pauling's contemporaries may have exceeded him in independent-mindedness, none of them displayed the same degree of wide-ranging inventiveness.

When asked by a student in the 1930s how he managed to originate so many ideas, Pauling replied that it was easy to come up with ideas, but not so easy to determine which were good ones. This characterization was unduly modest—Pauling's audacity was coupled with deep chemical intuition, and his overall ability to identify sound ideas and reject unsound ones was remarkable. Perhaps more than that of any other modern scientist, Pauling's work spanned all the levels of physical reality, from the submicroscopic world of elementary particles to the macroscopic world of living organisms.

Pauling worked on so many different subjects, even within the subdiscipline of molecular biology, that we will consider only a few key topics: Pauling's study of hemoglobin and sickle-cell anemia, his profound and fundamental work on the secondary structure of proteins, and his influential model of antibody formation.

While plants and animals are very complex, they all contain a limited number of molecular substances. Most living matter is composed of very large molecules. These molecules are usually formed into long chains with patterns that repeat themselves over and over. Some of the bonds between the atoms in these molecules are very strong. Many of these are formed by valence bonding similar to that which Pauling found in nonorganic molecules. Others, however, are much weaker and do not involve the sharing of electrons between the atoms. These electrostatic bonds occur because two atoms that are close to each other have small effects on each other's electrical charges, effects often referred to as the *van der Waals forces* after the Dutch physicist who first proposed them. These weak bonds are essential to living organisms since they can be broken down and reestablished fairly easily, making possible many of the chemical reactions that are fundamental to life.

These weak bonds often involve hydrogen atoms. Each hydrogen atom has only one electron and can therefore enter into a strong valence bond with only one other atom. Hydrogen atoms do, however, form weaker bonds with other atoms as well as the one strong valence bond. Because hydrogen atoms in water molecules form bonds quite freely, water is the essential medium in which biological processes take place. Carbon, oxygen, and nitrogen atoms are frequently linked with hydrogen atoms to form complex biological molecules, whose shape changes as the hydrogen bonds are broken and re-formed.

While the organic molecules are large and complex, involving hundreds or thousands of atoms, Pauling and other investigators thought that regular repeating patterns were the key to unraveling their structure. Pauling was convinced that his fundamental rules for atom-to-atom bonding would allow him to determine the structure of complex molecules. He often worked by building physical models, which incorporated his findings about the angles at which molecular bonds could be formed in complex molecules. Working with these models made it possible for him to see the relationships between atoms physically rather than having to represent them in complex mathematical formulas. This was particularly necessary for complex organic molecules, which were too complex to be described well in two-dimensional drawings.

Some scientists disparaged these models, which bore a striking resemblance to those that Pauling prepared to illustrate the structure of simpler compounds to his students. Solving structures in this way was too much like playing with puzzles or toys. However, the models made it possible to use the same skills to visualize spatial relationships in identifying the structure of molecules too complex to be understood in any other way.

Pauling's first venture into research on complex organic molecules was a study of the magnetic susceptibility of hemoglobin molecules that he suggested to one of his graduate students, Charles Theorell, in 1935. They were interested in checking the theory that the oxygen carried by the blood entered into a true chemical reaction with the iron in the hemoglobin molecules, changing its molecular structure. They found that hemoglobin that was carrying oxygen and hemoglobin that was not carrying oxygen reacted differently to a magnetic field. Pauling interpreted this finding as showing that the molecular bonds in the hemoglobin without oxygen are weaker, ionic bonds, whereas those in the molecule united with oxygen are stronger, covalent bonds. Finding this difference in the laboratory greatly excited Pauling's interest in identifying the structure of complex organic molecules, especially protein molecules, which are fundamental to hemoglobin and other vital components of the human body.

When Robert Corey joined the staff at Caltech in 1937, he and Pauling set to work on protein structure. This was a logical first step, and one that was being taken by many other scientists, including the English crystallographer Thomas Astbury, who began work on the topic several years before Pauling did. This change of focus was encouraged by the Rockefeller Foundation in its support of research on biological compounds. It was also a major focus of Britain's Cavendish Laboratories, which were headed by Sir Lawrence Bragg. Thus Pauling moved into a highly competitive field, again in the mainstream of scientific inquiry at the time.

It was also a very difficult field, particularly because the information that the scientists obtained from the X-ray diffraction pictures was blurry and incomplete. It gave certain information about the molecules, but often a large number of different structural models would fit the available information reasonably well. Constructing these models was difficult and time-consuming, and large gaps in the picture had to be filled in by the scientist's imagination. In the mid-1930s, Pauling was unable to untangle the structure of hemoglobin or other complex protein molecules, but he was convinced that previous models were incorrect. The major problem was the imprecision of the X-ray data.

In 1945, however, Pauling did make an important discovery concerning hemoglobin. He was seated at a dinner party next to a physician who was engaged in research on sickle-cell anemia. In this disease, the red blood cells are deformed when they are in the veins, but resume their original shape in the arteries. Pauling hypothesized that this sickling was due to a molecular defect in the hemoglobin itself that created difficulties in its attaching itself to oxygen molecules. He asked one of his associates, Harvey Itano, to study samples of blood from people with sickle-cell anemia and from people with the sickle-cell trait but without the disease, and to compare these samples to blood samples from individuals without the disease or the trait. It took several years of difficult research, but finally they were able to use an electrophoresis apparatus that had been recently improved and used by Arne Tiselius to separate the components of a mixture of proteins based on their different numbers of charged amino acids. Pauling and Itano showed that under certain conditions the normal hemoglobin had a negative charge while the sickle-cell hemoglobin had a positive charge. Their results were published in *Science* in 1949.

This was an exciting new idea, one often cited as a landmark in the history of biochemical genetics. Archibald Garrod, in 1902, had formulated the concept of inborn errors of metabolism, in discussing the human inherited biochemical disease alkaptonuria. Pauling's notion of molecular disease was a natural step along this path. An important disease, sickle-cell anemia, had been shown to correlate with an electrophoretic aberration in the hemoglobin molecule, which seemed certain to follow from a definite biochemical abnormality at the molecular level. A path was now clear for other researchers to search for other molecular diseases.

A genetic study by James V. Neel confirmed two months later that sickle-cell disease was hereditary, inherited in the Mendelian manner, which meant that individuals with one defective gene had the sickle-cell trait; those with two invariably had the disease. The disease developed in African populations because the sickle-cell trait provides protection against parasites that cause malaria. Research by Max Perutz indicated that when sickle-celled hemoglobin was not carrying oxygen, it stretched out in long filaments that made the cell itself deformed and rigid. Finally, the Cavendish biochemist Vernon Ingram was able to find the exact place where the defect occurred in the sickle-cell hemoglobin through the process of chromatography plus electrophoresis, which provides a unique "fingerprint" of a molecule. There is as yet no cure for sickle-cell anemia, but Pauling suggested a practical if impolitic solution. In his speeches, he suggested that carriers identified at birth be tattooed on the forehead with a distinctive symbol to warn them not to marry each other.

The work on hemoglobin was important and exciting, but Pauling's greatest triumph in chemical biology was his discovery and exploitation of the so-called secondary structure of proteins, the helical peptide-linked polymer molecule. This discovery forms the basis of modern molecular biology and ranks alongside the rules for mineral chemistry and the theory of the chemical bond among Pauling's premier scientific achievements.

It was already clear in the nineteenth century that an understanding of protein structure was crucial to the understanding of biological systems. But not until this century was it possible to study the structure of proteins in any detail. The first big step was Emil Fischer's characterization of proteins as *polypeptides*—long chains of amino acids, interlinked in various ways, and then folded up to form three-dimensional structures. Gradually more and more information became available from X-ray diffraction techniques and also the new technique of *paper chromatography analysis.* In the first years of the 1950s, a British scientist, Fred Sanger, and his colleagues published results using paper chromatography to elucidate the order of amino acids in protein chains—what is now called the *primary structure* of proteins.

University of London scientist J. E. Bernal recalls that, in the mid-1940s,

> It was clear that a successful attack on the complete protein structure could be made, but there were still many difficulties. . . . Two modes of attack suggested themselves: the first was a straightforward X-ray crystallographic approach. . . . The second was a model building method based on an exact knowledge of the structure of amino acids and smaller peptides themselves. . . . I remember very well discussing the problem with Pauling just before the war. He was in favor of the second method, which I thought indirect and liable to take a long time. Nevertheless, it was Pauling's ideas that were to have a decisive effect on the result.[1]

Pauling, once again, triumphed because of his ability to combine different sorts of information, and to infer structure by a combination of intuition and empirical investigation. Bernal's approach, to obtain the best data possible and follow them, was bogged down in the complexities of the problem. Pauling's approach was to guess the correct structures and check his answers by using X-ray diffraction. This same approach had been effective in his early studies of mineral chemistry and his theory of the chemical bond. In each case, he used simple but imaginative ideas about geometrical and physicochemical properties to narrow down the "search space" of possible structures.

Pauling's theory relied on three crucial insights: two regarding the details of particular chemical bonds, and one regarding the general nature of the chemical structures underlying living systems. His first insight was related to the X-ray analysis of the dipeptides and tripeptides (groups of two and three peptides, respectively), in particular to the bond between a carbon atom and the succeeding amide nitrogen atom. The general view was that this had to be a simple, single bond, capable of any orientation as if the two parts were held together by string. But Pauling, inspired by his valence bond theory, hypothesized a quantum "resonance" between this bond and the neighboring carboxyl group, a resonance that prohibited rotation about the bond. This hypothesis, which arose directly from his theory of the hybridization of orbitals, was later shown to be true and greatly simplified the problem of deducing protein structure by X-ray diffraction. It narrowed the search space; one no longer had to consider every possible orientation for this crucial bond, only one.

Having had this insight, Pauling wanted to propose a helical structure for polypeptide chains, based on the nonrotating bond he had discovered (figure 12). At first sight this was a noncontroversial idea: helical structures for polypeptide chains had been suggested before. But in order to make the details work, radical modification of the basic ideas of crystallography was required. It had been thought, up to this point, that helical structures could occur only in crystals with two-, three-, four-, or sixfold symmetry, with the same patterns repeating over and over. But imposing this rigid mathematical symmetry on large biological molecules was unacceptable—it would not leave sufficient freedom. Pauling's idea, which Bernal called a "stroke of genius,"[2] was to discard the concept of symmetry altogether. Instead of a repeating sequence of peptides, why not a nonrepeating sequence? Why not an entirely irrational sequence, a sequence that encoded information in some way other than simple repetition? The crystallographic restrictions on structure were really intended to hold between molecules, not within molecules, anyway. With this idea, in one stroke, Pauling set science on the course toward modern molecular biology. Today we know a great deal about the ways information is encoded in nonperiodic structures on polypeptide chains, structures not consisting of the same thing over and over again. At the time, however, no such knowledge was available.

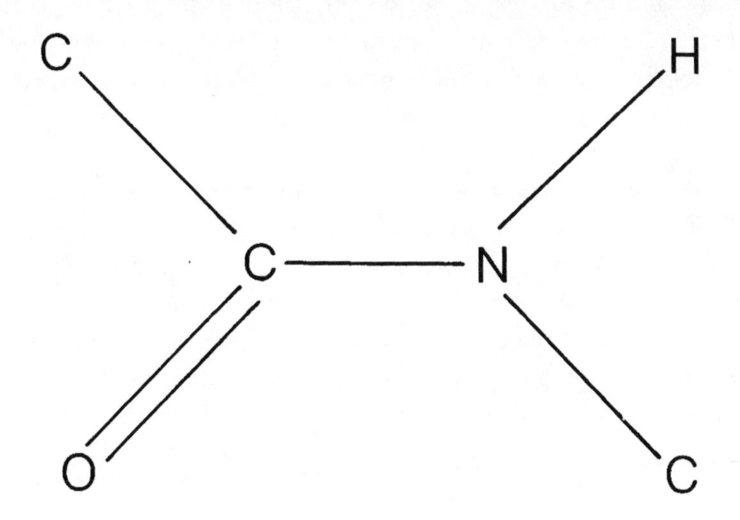

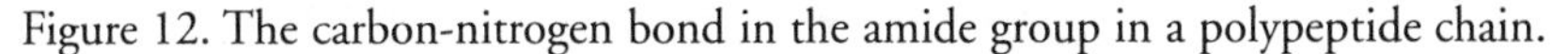

Figure 12. The carbon-nitrogen bond in the amide group in a polypeptide chain.

And nonperiodicity led Pauling readily to his third key insight, a technical theory of the stabilization of the helical structure. The stability of the structure, he proposed, was due to hydrogen bonds between the oxygen of the carboxyl group and the nitrogen of the amide group in the same chain. All these hypotheses, put together, led to a simple, concrete model of proteins as helices—the *secondary structure* of proteins, as opposed to the *primary structure,* which is the particular sequence of amino acids or other peptides.

One might speculate that Pauling would have worked more productively at this time if he had not been so involved in traveling and lecturing. But the matter is not so clear. Pauling said that solutions to problems often came to him at unexpected times, often while his conscious mind was not focused on the topic in question. He reported that this was true of perhaps his best known single discovery, the alpha-helix structure of the polypeptide chain. According to Pauling, the idea for this structure occurred not in his laboratory but during his trip to London in 1948. This is how he described the discovery on the television program "Nova" in 1977:

> I had a cold in the spring of 1948 in Oxford where I was Eastman Professor at Oxford. And I stayed in bed for two or three days, three days perhaps, and the first day I read detective stories and just tried to keep from feeling miserable and the second day too, but I got bored with that, so I thought, why don't I think about the structure of proteins. Eleven years earlier I had worked one summer with molecular models of the sort, the string and the rod and ball molecular models, and with my ideas about the structure of compounds of that sort, that I thought would give the right answer, and I had failed. Well I tried again—I didn't have any molecular models with me in Oxford but I took a sheet of paper and sketched the atoms with the bonds between them and folded the paper to bend one bond at the right angle, what I thought it should be relative to the other, and kept doing this, making a helix, until I could form hydrogen bonds between one turn of the helix and the next turn of the helix, and it only took a few hours of doing that to discover the alpha helix.

After Pauling recovered, he visited Bragg's laboratory at Cavendish. At the time there were two leading laboratories in protein research,

Bragg's and Pauling's. There was rivalry as well as cooperation between them. Bragg, Kendrew, and Perutz were the top scientists at Cavendish, and when Pauling visited their laboratory they showed him a tentative model of a hemoglobin molecule. Pauling thought he saw an error in it but didn't say anything. Pauling said he didn't mention the alpha-helix idea to them but waited until he got back to his laboratory in Pasadena, where he could check it out. Meanwhile, Bragg, Perutz, and Kendrew published their research in the *Proceedings of the Royal Society.*

Their article was an embarrassing failure. Pauling said that it failed to take into account basic chemical information that they could have found in his book, *The Nature of the Chemical Bond.* Francis Crick had told his superiors at Cavendish that they were on the wrong track, but he was a doctoral student with little status at the time and they ignored him. Bragg was annoyed by Crick's mannerisms and was eager to be rid of him. Bragg was so humiliated by the episode that he never again put his name on a paper for publication.

Meanwhile, back in Pasadena, Pauling invited a bright young scientist, Herman Branson, to become involved in the research on protein molecules. Branson was spending the academic year 1948–49 at Caltech as a senior fellow of the National Research Council. Branson's account of the discovery of the alpha helix in a 1984 letter is quite different from Pauling's:

> Pauling graciously proposed that I look at how the amino acids might be arranged in a protein molecule following the bond lengths and angles that had been established for amino acids. I worked on this problem in the math library, reading all I could find on projective geometry, odd equations, and the like. At the end of two or three months, I had found two spiral structures which fit all the data. They were the alpha and gamma helices. I took my work to Pauling who told me that he thought they were too tight, that he thought that a protein molecule should have a much larger radius so that water molecules could fit down inside and cause the protein to swell. I went back and worked unsuccessfully to find such a structure.
>
> In the meantime I had made models. One day Corey came by with some large F-H atomic models. We attached them as the alpha helix. Corey's response on viewing the spiral was "Well, I'll be damned."

> I wrote up what I had done and gave the paper to Pauling when I left in the Summer of 1949. I heard nothing until I got this letter from Pauling dated October 6, 1950. I interpreted this letter as establishing that the alpha and gamma in my paper were correct and that the subsequent work done was cleaning up or verifying. The differences were nil. Pauling had proposed the problem and I certainly would never have come near it without his invitation.[3]

Meanwhile, Pauling and Corey sent a short note to the *Journal of the American Chemical Society,* published in November 1950, titled "Two Hydrogen-Bonded Spiral Configurations of the Polypeptide Chain." This short paper merely announced the findings without supplying the details. Pauling then wrote his letter of October 6, 1950, to Branson, who by that time was at Howard University. In it, Pauling said:

> I enclose a draft of a manuscript which we propose to publish in the *Proceedings of the National Academy of Sciences.* Dr. Weinbaum did a great amount of work, based on your original notes, and I think that a discussion of the configurations of the spirals is in good shape.
>
> Please let me know if there are any suggestions that you want to make about the wording of the paper. Also let us know if it is satisfactory to you to publish it in the *Proceedings of the National Academy.*

The paper, "The Structure of Proteins: Two Hydrogen-Bonded Helical Configurations of the Polypeptide Chain," was published in the *Proceedings of the National Academy of Sciences* in 1951 with Pauling as the senior author, Corey as second, and Branson as third. These two helical structures, which they called the alpha and gamma helices (figure 13), constituted an important scientific advance.

This discovery was part of an ongoing research program directed by Pauling and Corey. The article on the alpha and gamma helices was followed by seven other articles by Pauling and Corey in the *Proceedings* in April and May 1951, which described their findings about a variety of structures of proteins in detail. The articles were especially applauded by American scientists, who were sympathetic with Pauling's political troubles and proud to see an American succeed where English scientists had failed.

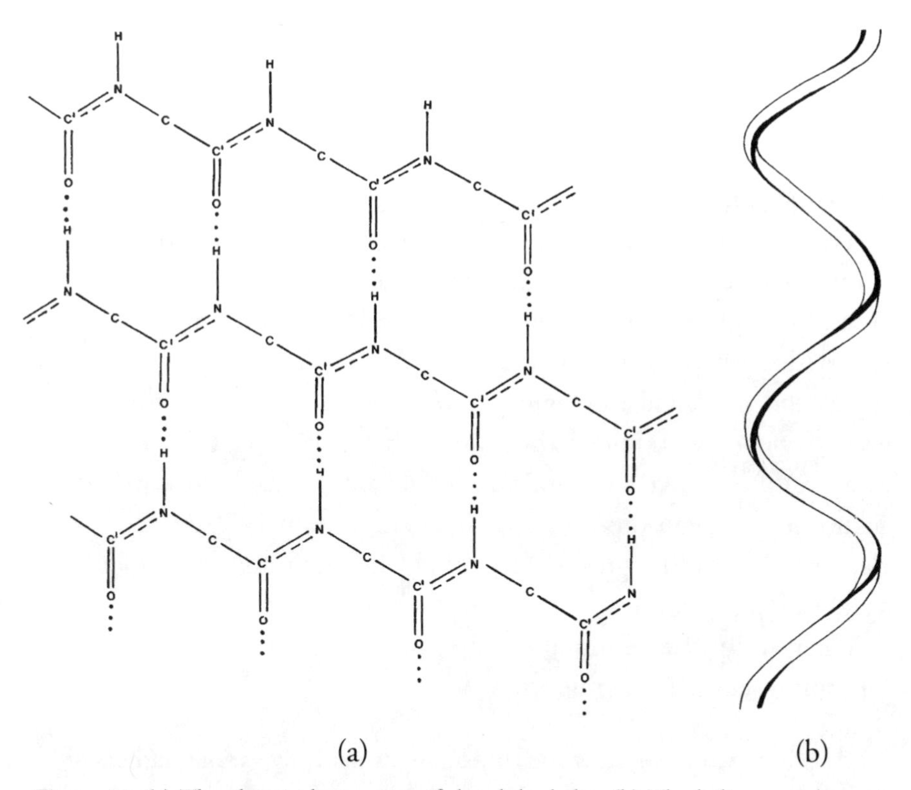

Figure 13. (a) The chemical structure of the alpha helix. (b) The helix, a mathematical structure that is ubiquitous in molecular biology.

Herman Branson, who went on to become President of Lincoln University in Pennsylvania, believed that he had not received sufficient credit for the discovery of the alpha and gamma helices. He was particularly angry when a friend sent him a clipping from the *Pasadena Star-News* with the headline "Secrets of Proteins Uncovered: Two Caltech Chemists Untangle Building Blocks of Life." The two chemists highlighted were Pauling and Corey; Branson was mentioned only in a brief paragraph, which stated, "Dr. Herman Branson, Howard University professor, assessed Dr. Pauling's work and found yet another helical, spring-like arrangement of atoms which fitted the many conditions of proteins."

In his 1984 letter, Branson told us that he "truly resented" this paragraph, which "seemed to be the 'official' version." At the same time, Branson assured us that Pauling was "one of the impressive scientific intellects of our age who deserves the Nobel Prizes," and he expressed reluctance to detract from the "twilight of his life." Interviewed shortly before his death in 1995, he stood behind his 1984 letter and further asserted that Corey had had nothing to do with the discovery of the alpha and gamma helices.

The issue of credit for scientific discoveries is a difficult one, and it is common practice for senior professors to take the major part of the credit for discoveries made by younger researchers working under their supervision on projects that they have designed. Memory can also be tricky, and similar ideas may occur at about the same time to two scientists who are working on the same problem. Pauling did list Branson as a coauthor on the scientific paper detailing the discovery, and he acknowledged Branson's assistance in his important 1970 *Daedalus* article, "Fifty Years of Progress in Structural Chemistry and Molecular Biology," Pauling's definitive account of the historical record.

When we queried Pauling on the general subject of assigning credit for joint discoveries, he told us:

> My policy was to consider the contributions made by various authors of a paper, including the suggestion that the work be done, and to assign positions in the list of authors in accordance with the contributions that were made by them.
>
> Many researches were suggested by me to my students or other collaborators, and in some cases the fact that I had originated the work seems to me to be important enough to justify having my name as first author.[4]

Pauling's recollection of the discovery of the alpha helix has been supported by Oregon State University Chemistry professor David Shoemaker, who remembers seeing helical models of paper in Pauling's apartment at Oxford in early 1948. Pauling's daughter, Linda, also remembers seeing these models. Perhaps more importantly, Pauling, Corey, and other scientists had published several articles in scientific journals that, in the judgment of Harvard chemist William Lipscomb, "formed the basis for the correct structure of the alpha and gamma helices." Lipscomb was at Caltech from 1941 to 1946 and remembers that these studies were well known to him and others at the time. In

Lipscomb's judgment, "the comments of Branson should be viewed in a much larger context, in which the results of the actual model building proceed inevitably from the information available, whether or not Pauling told Branson of the March 1948 results obtained by Pauling in Oxford."[5]

The alpha and gamma helices were recognized as brilliant models, but it was not clear how common they were in the real world. Some of Pauling's attempts to apply these and other protein models turned out to be incorrect. For instance, in his study of globular proteins, Pauling assumed that the structure was one of rods of polypeptides arranged parallel to each other in different orders. But Francis Crick demonstrated that this was incompatible with the observed X-ray data. As it turned out, the structure was not a simple one at all, and there was no way that Pauling could have guessed it at the time. Eventually, scientists found that the alpha helix occurred only in certain proteins rather than in all globular proteins, as Pauling had thought. The "pleated sheet" model that Pauling and Corey thought would be found in stretched hair and other similar fibers has been shown to exist but less frequently than they postulated. Their paper on the structure of collagen was shown to be completely wrong.

Fortunately for Pauling and the other scientists involved in this story, the successful findings that scientists make are remembered and the errors are graciously overlooked by their colleagues. Certainly the English scientists, who themselves had just published erroneous models of the proteins that Pauling, Corey, and Branson had correctly described as an alpha helix, could not criticize Pauling's errors. Indeed, Perutz published the first independent finding confirming the alpha helix.

It may seem, from these oversights, that Pauling's geometric intuition was not quite so useful in the biological realm as it was in the mineral realm. In many cases, he tended to overestimate the degree of regularity of biological structures. Yet, it was he who originated the idea of nonrepeating helical structures, which lies at the heart of modern molecular biology. Although his success percentage was definitely lower in chemical biology than in mineral chemistry, his knowledge, intuition, and analytical ability were sufficient to lead him to many discoveries in molecular biology.

• • •

The helical structure of proteins was recorded as Pauling's major biological triumph, solidifying his position as a master of all levels of chemical structure, from the simplest minerals up to the most complex organic molecules. But Pauling also did important work in a great number of other areas of chemical biology. His work on the immune system was particularly influential. Pauling's model of antibody formation, published in 1940, was not all that different from earlier models—but because of his ability to explain his ideas and intuition about the fine points of chemical structure the model had wide impact on immunochemistry. Immunology has now moved on to different ideas, but Pauling's early investigations were an important part of the process of getting the science where it is today.

Antigens are the bad guys of immunology—germs or viruses that invade the body and make it ill. *Antibodies,* special folded-up proteins, are the good guys, which kill the antigens and allow the body to restore itself to health. In the early part of this century, the process of antibody formation was a matter of some debate. Early theories by Paul Ehrlich and Hans Buchner were discarded in favor of a "template theory," based on the metaphor of the lock and key. The antigen is the lock, and in order for the antibody to kill the antigen, its combining sites must have a shape complementary to that of the antigen, just as the shape of a key is complementary to the shape of the lock into which it fits. The idea was that the foreign antigen somehow affected the folding of the protein molecule, in such a way that the final antibody molecules contained areas that "matched" the foreign antigen.[6]

What was unknown, however, was whether the shapes of different antibodies' combining sites were determined by differences in amino acid sequences, or just by differences in the final folding. Pauling came out on the side of differences in final folding. He postulated that each antibody had two combining sites, and that the specificity of combination with antigen was the result of the antigen's influence on the final folding. His theory was presented in simple diagrams, like figure 14, which shows the antibody molecule forming as a result of interaction between the globulin polypeptide chain and the antigen molecule.

Today, as a result of the efforts of such theorists as Frank Burnet and Niels Kaj Jerne, much more is understood about the immune system.[7]

Figure 14. Pauling's multistage model of antibody formation. (*left*) The six stages in the formation of an antibody molecule as the result of interaction of a polypeptide chain with an antigen molecule (represented as a black mass). (*right*) The antibodies surrounding an antigen molecule and thus inhibiting it from further antibody formation.

The most common view today is that the antigen does not directly affect antibody formation: instead, the immune system already contains a sufficient variety of antibodies, and the appropriate antibody to match an antigen is found by a process of "natural selection," according to which antibodies that match better are more frequently replicated. This view is surprisingly reminiscent of Ehrlich's early speculations, and it contradicts Pauling's ideas. Thus, it seems, Pauling's elegant diagrams were almost certainly incorrect. But his ideas led to several experiments, many of which were classics in immunology. Later on, he was to note a similarity between his work on antibodies and the double-helix model of deoxyribonucleic acid (DNA): both make essential use of the phenomenon of "matching" or complementarity. This is a very instructive lesson in the workings of science: it is important for a theory to be right, but it is just as important for a theory to be suggestive, to give rise to further avenues for investigation.

• • •

Throughout the 1920s and 1930s, Pauling, like most scientists, devoted most of his time and energy to his work. He and Ava Helen had little time to visit relatives in Oregon, and they gradually grew apart from their families there. Their children grew up knowing little about their parents' background before Pasadena. Their fourth and last child, Edward Crellin Pauling, was born in 1937, when their first child, Linus Jr., was entering adolescence. Peter Jeffress Pauling, named after Linus's boyhood friend, was seven at this time, and his sister, Linda, was six.

The three older children attended a rigorous private school. Linus had little time to spend with them, although he was always there in emergencies. On one occasion when he took Peter and Linda with him to the laboratory, an electric heater ignited Linda's dress. Linda remembers, "I was standing too close to it because he told me to be careful and not stand too close or my dress might catch on fire. It did." Pauling beat out the flames with his bare hands before she had time to be frightened or burned. He didn't nag about small matters, and when one of the children misbehaved he took a rather detached and pedantic attitude toward them, sitting down with the errant child and itemizing, step by step, the actions needed to correct the mistake. He then listed the steps to be taken to correct the situation. He never asked for nor heard their defense of themselves or speculated about why they had transgressed.

The Pauling children lived in an environment that was stimulating and noncoercive, but somewhat sterile emotionally. They knew that their father was a genius, and that others held him in awe. On one occasion, when visiting their school on career day, he dazzled the students by reciting the periodic table of elements from memory.

In 1939, the Paulings moved into a unique home that they had built on a large lot in the hills overlooking Pasadena with wings out from the living room at 109° angles to approximate the angle between the carbon-carbon bonds in many organic molecules. The house was cluttered with toys and books, crayons and clay, bathing suits and skates, balls, records, collections, scrapbooks, and models of molecules. The children were also exposed to music and practiced various musical instruments.

In the traditional role relationships of the Pauling family, Ava Helen was primarily responsible for child care and household chores. This corresponded to the way Linus and Ava Helen had been brought up, and was the norm in the community at the time. Somewhat unusual, however, was the extent to which Linus's career took precedence over family demands and monopolized Ava Helen's attention. It was always clear to the children that their father's career came first and that their chances of matching his achievements were slight. In moments of exasperation, Ava Helen would sometimes chide her sons, "You'll never be the man your father is."

During this period, though largely absorbed in his work, Linus did participate in leisuretime activities that were relaxing and undemanding such as swimming or hiking, and he accompanied his wife to concerts and necessary social events. He was not active in community or school affairs, nor did he take any public political positions. He wrote no letters to the editor and avoided all activities in which he might have to sit on a committee.

The depression had no major impact on Linus and Ava Helen because his job at California Institute of Technology was secure. Of course, he could hardly fail to notice the stock market crash of 1929, or the first few years of the great depression. And although he strongly opposed it, his salary, like that of all other faculty, was cut by 10 percent in 1932 by President Millikan because of financial exigencies caused by the depression. Fellowships to graduate students were also cut, but Caltech faculty and students were still better off than many of their fellow citizens.

Pauling's mother had supported Woodrow Wilson in 1912, but when it came time to vote in 1932, Pauling followed the usual family tradition and voted Republican. As he explained later,

> Until the 1930s I had essentially no interest in politics or in other broad questions, because I was so enthusiastic about my work in chemistry and X-ray crystallography. My parents were Republican, and I did not think much about political matters when I continued to vote Republican in 1932. Then I became interested enough to form some opinions of my own, which resulted in my support of Upton Sinclair.

When Pauling changed his registration from Republican to Democrat in the summer of 1934, he committed himself to a political philosophy that remained central to his thinking for the rest of his life. Upton

Sinclair's campaign of 1934 was an exciting time in California politics; even today some refer to it as "the campaign of the century." Sinclair, a famous radical novelist, lived in Pasadena, and Pauling visited him in his home there. Albert Einstein also met Sinclair in Pasadena, but Pauling had no recollection of having discussed Sinclair with Einstein. Sinclair had been a candidate of the Socialist party for several offices before 1934, winning only about fifty thousand votes. In 1934, some friends suggested that he change his registration to Democrat and run in the primary for governor. He accepted the idea, much to the dismay of Norman Thomas and others who remained loyal to the Socialist label.

Although Sinclair abandoned *socialism* as a word, he continued to advocate many socialist ideas in more acceptable language. His slogan was "End Poverty in California" (EPIC), which he proposed to do by instituting a system in which "the means of producing and distributing the necessities of life should be in the hands of the entire people, to be used for the people's equal benefit, and not for any privileged class." To those who accused him of promoting socialism, he simply said that he was interested in solving urgent problems, not debating theory.

Sinclair's practical ideas were innovative. Instead of nationalizing big businesses, as socialists advocated, he proposed setting up a parallel economy that would follow the slogan "Production for Use." The state would take over failed businesses and farms and pay workers in scrip they would use to purchase food and other necessities produced by other workers working on the same plan. With time, he expected that these cooperative companies would drive capitalist enterprises out of business, and the whole economy would be organized around the concept of production for use.

Sinclair won the Democratic primary, which was the largest victory of any candidate with socialist ideas in American history. He hoped to get the endorsement of Franklin Roosevelt, the official Democratic party candidate. Roosevelt, however, stayed on the sidelines when the conservative forces in California rallied to attack Sinclair for his radical ideas. The *Los Angeles Times,* in particular, led a virulent media campaign against him, often featuring selected quotes from Sinclair's novels and other writings. To a degree, Sinclair fell into their trap by advocating a state motion picture company that would compete with the private studios that were California's leading industry. The studios mounted a major newsreel campaign against him, suggesting that he was the candidate of the poor, the immigrant, and unwashed against the American middle class. There was also pressure against faculty at California Institute of Technology who were sympathetic to Sinclair, but Pauling did not recall being under any pressure himself or having any knowledge of faculty who feared reprisals for supporting Sinclair.

Sinclair was an advocate for the common man against the powerful interests. He was defeated in the first electoral campaign in which the establishment used the mass media to manipulate public opinion. Pauling's views were shaped by this experience. After 1934, he saw politics as a struggle of the common people against the establishment. He was acutely sensitive to the importance of the mass media and learned how to use them effectively. He also learned the importance of not being labeled as an extremist or radical and immediately went on the offensive when the mass media attacked him as a Communist sympathizer in the 1950s and 1960s.

Pauling voted for Roosevelt in 1936, as did most of Sinclair's supporters, and continued to be a liberal Democrat for the rest of his life. His first active involvement in politics was not a response to the depression or Roosevelt's domestic agenda, however, but a reaction to the rise of fascism in Europe and the threat of world war.

Many of the Pauling's European friends' lives were disrupted by the emerging power of Adolph Hitler. Scarcely a week passed that did not bring some bitter news about an old friend or acquaintance. Goudsmit was safe as a professor at the University of Michigan, but both his parents were killed by the Nazis. Einstein also left Europe early and was safe at Princeton, while Max Born and his wife fled to Great Britain, and Schroedinger went to Ireland. Debye left his position as director of the Kaiser Wilhelm Institute for Physics in Berlin rather than give up his Dutch citizenship to become a German. He was lecturing at Cornell University when Hitler invaded Holland and was invited to stay on.

In 1940 Niels Bohr gave up his scientific work to "expound the cause of liberty" in his own country. The Nazis invaded and sacked his laboratory, only to find that he had left no useful data behind. He was flown to England in a Mosquito bomber and eventually did atomic research at

Los Alamos. Jewish scientists had little choice but to leave Germany, but others had hard decisions to make. Werner Heisenberg became the top scientist under the Nazi regime, while Max Planck remained in Germany as a dissenter. His prestige protected him from death or imprisonment, but his second son was executed for being part of a group that plotted to assassinate Hitler.

One of Pauling's regrets was that he did not take a stand against anti-Semitic hiring practices within academia in the late 1930s. Pauling often received letters from other universities saying that they wanted to hire a crystallographer or a theoretician, but that they did not want to hire a Jew. Pauling did not protest this practice at the time, although he did not follow it in his own hiring. He later told science historian Robert Paradowski that he was ashamed that he did not speak out on the issue at the time.

Although Pauling did not act against anti-Semitism in academia, he and Ava Helen were morally outraged by the events in Europe and wanted to do something about them. But what should they do? Many citizens were inactive simply because they didn't know what they could do about problems in Europe. Linus Pauling differed from these citizens not so much in his concern for the world's problems, as in his confidence that he could readily find solutions for them. His response to the world's wrongdoings was similar to his treatment of his children's misbehavior. He assumed that his fellow citizens were well intentioned and that they would take action to correct the problems if they understood what needed to be done. His role was to analyze the problems, find solutions, and then educate the public about them.

Pauling was not a social scientist, however, and he did not intend to become one. Instead, he sought answers in the writings of others who shared his values and had already studied the problems. In 1940, he found the solution to the problem of Nazi Germany in the writings of a New York Times reporter, Clarence Streit, whose first book, *Union Now,* was a best-seller in many languages. Streit proposed a federal union of all the English-speaking democracies, particularly England, Canada, and the United States, a federation that would then impose economic sanctions against any aggressor nation.

Pauling's first important lecture on a nonscientific topic was an address to the Pasadena Federal Union Club on July 24, 1940. His message was a summary of Streit's arguments. The war in Europe was a struggle between democracy and autocracy, and the United States would be placed under economic vassalage to Germany if England were defeated. He recognized that the Union Now proposal would be, in effect, a declaration of war against Germany but argued that it would be better to fight now than later after England was defeated. Possibly war could be prevented through an economic boycott of Nazi Germany.

The proposal for a political union with Great Britain was never seriously considered by the political leaders of either country. The underlying idea of greater American support for the British war effort, however, did gain widespread acceptance, even before the bombing of Pearl Harbor.

In March 1941, while giving a speech in Boston, Pauling noticed that his face was all puffed up. He went on to New York, where he received the William Nichols medal from the New York Section of the American Chemical Society, in recognition of his work on the chemical bond. Physician friends asked to examine him and, after running some tests, diagnosed glomerulonephritis, a kidney disease that produces facial edema and hypertension. They recommended that he return to California immediately and consult a specialist, Dr. Thomas Addis.

Fortunately, Pauling took their advice even though it meant canceling a trip to the Mayo Clinic, where he was scheduled to speak. Dr. Addis put him on a regimen that included long weeks in bed; a strict diet including no fat, salt, or sugar; and vitamins and liver extracts. Addis thought that, if Pauling had sought treatment while visiting the Mayo Clinic, the doctors would have followed a more conventional medical treatment with polysaccharides, chemicals such as starches which include many units of sugar. This, in Addis's opinion, would have relieved the facial edema without curing the underlying problem, which would have led to Pauling's death. At the time, Pauling was unaware that his grandfather, Linus Wilson Darling, whom he resembled in many ways, had been suffering from nephritis when he died of a heart attack at the age of fifty-five.

Pauling recovered slowly under Ava Helen's careful supervision of Addis's prescription. This personal experience gave Pauling good reason to be sympathetic to natural treatments aimed at restoring the body's chemical balance, as well as the value of attentive home care.

Linus felt lonely traveling anywhere without Ava Helen, so she accompanied him on almost all his trips despite the responsibilities of raising four children. With Linus's illness, Ava Helen became his nurse as well, watching over his health and his diet almost as if he were an infant unable to care for himself. She gave Linus her total devotion and unquestioning support, something his mother had been unable to give him after his father's death. Ava Helen found this role demanding, but she felt that it was necessary to nurture his genius. When an interviewer for *Cosmopolitan* asked her what it was like to be married to a genius, she explained:

> The most difficult part for me was assuming total responsibility for all the details of ordinary living. I have to do all the routine things, like having the cesspool pumped out. For fifteen years my husband was on a protein-free diet—no meat, no fish, no chicken. I had to weigh his food to find out how many grams of protein each item of food contained. It was a great deal of work, but he got better and was able to keep on working.[8]

The Pearl Harbor attack and America's declaration of war against Germany and Japan made the Union Now movement academic. Pauling's laboratory, like many others, focused primarily on military research during the war years, and Pauling served on the explosives division of the National Defense Research Commission, the consultative committee on medical research of the Office of Scientific Research and Development, and as a member of the Research Board for National Security. While Pauling had been a few years too young to serve in World War I and was too old for military service in World War II, he supported U.S. policy in both of these conflicts.

He continued to teach classes at Caltech, but the mood was gloomy as students enlisted or were drafted and reports of casualties flowed regularly into the campus community. Linus Jr. persuaded his parents to permit him to join the Air Corps even though he was not yet eighteen. They continued to receive reports of death, torture, or exile inflicted on European scientists by the Nazis. Pauling felt that his best role in the war effort was as a scientist and largely put aside his theoretical studies to engage in research for the war effort.

At a meeting of the chiefs of chemistry laboratories in Washington, the scientists were told of an urgent need for an oxygen meter for use on submarines and airplanes. Pauling drew up the plans for such a meter during his three-day train trip back to Pasadena. One week later, he started back to Washington with a working model of the meter. As he lay in his berth during the first night of the trip, he awakened and thought of his invention. To his consternation, he found that it was showing a reading that was too low and decided to get off the train as soon as possible and return to Pasadena. Fortunately, however, he soon looked out the window and discovered that the train was passing over the highest point in the Rockies, at an altitude that accounted for the low atmospheric pressure. The oxygen meter was widely used not only in the military, but in medical and commercial applications.

Pauling's laboratories also worked on explosives and medical problems. After three years of experimentation, Pauling and Daniel Campbell developed a synthetic blood plasma that could in some cases be used instead of real plasma. The entire family helped in the laboratory during the war when staff was short. Ava Helen worked as a chemical assistant on an experimental project aimed at extracting rubber from the guayule plant.

Pauling's first experience as a dissenter occurred unexpectedly when a worker from a war relief agency called and asked whether Ava Helen would employ as a temporary worker one of the two Japanese-American soldiers who had come to California from the Heart Mountain Relocation Center in Wyoming and were waiting to be called to Camp Shelby.

The two young soldiers wanted to see their old homes in California before they went to the front. Although these young American citizens had been summarily evacuated from their homes in the early days of the war because of their ancestry, they were still homesick for California. George Minski, the young man employed by the Paulings, had been a laborer on his father's farm near Glendale, California, before the evacuation, and the Paulings were glad to have his help in the yard.

On the morning of March 7, 1945, Peter Pauling came running into the house with the news that there were strange pictures painted on the garage. During the night someone had painted a great crude Japanese flag bearing the rising sun and the legend, "Americans die, but we love Japs. Japs working here!" on the garage. The mailbox had "Jap" painted on it. Crude and threatening anonymous letters arrived in the Paulings' morning mail. The children were angry and bewildered by the inexplicable behavior of the night visitors. The Pasadena family that had taken in the other young soldier was also visited in the night by vandals.

At this stage of the war, many people had already begun to be ashamed of the hasty evacuation of 100,000 Japanese-Americans from the West Coast. War hysteria, the greed of speculators hoping to purchase the Japanese-Americans' property at bargain prices, together with pressures from racist groups had made the evacuation possible on the West Coast, although never from Hawaii, where Americans of Japanese descent were much more numerous and closer to Japan.

The newspapers, on the whole, were critical of the night callers. They drew attention to the fact that Linus Pauling, Jr., had been in the Air Corps for eighteen months although he was not yet nineteen. Linus had not intended to make a public statement on this issue; his wife had simply agreed to help out a young man in need of work. But when the issue was thrust on him he responded forcefully, comparing the action of the vandals to the actions of Germans bent on persecuting Jews. He announced his firm intention to keep the young gardener in his employ until he was called to Camp Shelby, where he would be prepared for the battlefield in Italy, where many Japanese-American soldiers were suffering heavy casualties. In the relocation centers in the western United States the bells were sounded almost nightly to call the residents to the meeting hall to hear who was newly dead.

When the war ended, Linus Pauling, Jr., was among the soldiers who returned to resume their interrupted schooling. For his father, the end of the war meant resuming his prewar schedule. There were many demands on his time, particularly from organizers of scientific conferences around the world. Frequently, the lure was the offer of an award or honorary degree, or a request to serve as an officer or member of a prestigious scientific body. Accepting these offers placed significant demands on Pauling's time and energy, but he enjoyed traveling and speaking to interested audiences and found it hard to turn down an opportunity to do so.

Before the war, he had received honorary degrees from his alma mater, Oregon Agricultural College, in 1933, and from the University of Chicago, in 1941, as well as the Nichols Medal from the New York Section of the American Chemical Society in 1940. In 1947, he received the Theodore William Richards Award from the Northeast Section of the American Chemical Society, and an honorary degree from Cambridge University. Duties at home made it impossible for him to be in London in December 1947 to receive the Davy Medal from the Royal Society of London, England's oldest and most illustrious scientific organization, but in July 1948 he stopped in London to speak at the Centenary of the Chemical Society of the United Kingdom and to receive an honorary degree from the University of London. He was an honorary member of the Indian Academy of Science, a corresponding member of the Academy of Sciences of the Institute of France, a fellow of Britain's Royal Society, an honorary fellow of the Chemical Society of London, and a member of the U.S. National Academy of Sciences.

In early October 1948, Pauling received from the U.S. Army and Navy the highest tribute the military establishment can give to a civilian, the Medal of Merit. The medal was presented to him and four other Caltech faculty members at a handsome affair staged in Dabney Gardens. The faculty, student body, trustees, associates, and general public were invited to attend the solemn convocation. The 360th Army Band played resounding martial airs. Pauling's work with explosives, his oxygen meter, and his invention of a synthetic blood plasma were cited as reasons for the award.

In late November 1948, Pauling visited Paris to receive an honorary degree from the University of Paris. After the French celebration, Pauling had intended to stop off in London to attend a meeting of the Royal Society where he was scheduled to speak. He was especially eager to be there since he had been unable to accept the society's Davy Award in person, but London was fogged in and he missed his scheduled address. Instead, he went directly to Washington, D.C., where he attended a board meeting of the American Chemical Society, and on to Indiana, where he gave several lectures. His schedule was tight, he was often suffering from air sickness, and the weather was miserable.

In December 1947, Pauling had learned that he had been chosen as president-elect of the American Chemical Society (ACS). He was on his way to England to lecture at Oxford for the winter and spring terms, but on his return he took his responsibilities to the fifty thousand members of the society very seriously. He scheduled himself to visit over thirty ACS sections in Wyoming, Colorado, Oregon, Washington, Idaho, and Northern California. Many of these sections had never met a scientist of Pauling's reputation, and they were especially pleased and surprised that Ava Helen honored them by coming with her husband.

As head of the ACS, Pauling could not avoid the issue of civil liberties and anti-Communism, which was plaguing the academic world at the time. At universities all over the country, professors were being driven from their jobs on grounds that they were Communists or "Communist sympathizers." Often, all it took was an anonymous accusation or the suspicion of a congressional committee to finish a man's career. At Caltech outstanding scientists such as Sidney Weinbaum and H. S. Tsien lost their jobs. Weinbaum served a term in federal prison and gave up his scientific career, while Tsien had to leave the country to find employment.

Weinbaum was an emigrant from Soviet Russia, but he sympathized with the Soviet system and was a local leader in the movement to have the United States recognize the Soviet Union. Pauling had known him as a student and offered him a job as a research assistant. Weinbaum was especially strong in mathematics and he remembered that "anything that came up that involved mathematics in Pauling's work, I usually got a hold of it."[9] Later on, Tsien got him a job at the Jet Propulsion Laboratory at Caltech. In 1949, Weinbaum was asked to sign a new job application with a loyalty oath, and this was used as the basis for his perjury conviction. Pauling was supportive the whole time, telling him, "Don't worry, it will all be settled." Pauling also gave him a letter of recommendation, although Weinbaum remembered that Pauling said he preferred not to appear at his trial.

Weinbaum remembered, "It was very nice to work with Pauling. Pauling was very evenhanded with all the people, whether they were very bright or were not that bright. He didn't show any preferences in his relationships with people."[10] Pauling was often somewhat formal and distant with his assistants and employees, perhaps because of his concern to be evenhanded. He even addressed his secretary of many years, Dorothy Munro, as Mrs. Munro—never as Dorothy. However, he and Ava Helen became quite friendly with Weinbaum and his wife because of their European background and shared cultural interests. Weinbaum, and his political associate Frank Oppenheimer, say that Pauling was interested in political affairs at this time, but he was not active and was not a member of any organizations.[11] The Pauling's social relationship with Weinbaum cooled after he left prison because Ava Helen was angry with him for leaving his wife for another woman.

Although Pauling had not been active politically at Caltech, as president of the Chemical Society he rose to the occasion and took a forceful and courageous stand against McCarthyism at a time when there was no guarantee that his own career might not be in jeopardy. At the national meeting of the ACS in St. Louis in September 1948, when he was sworn in as president-elect, he spoke out strongly in criticism of the House Un-American Activities Committee's "scorn of humane and considerate treatment during its loyalty investigations." Pauling was especially concerned about the effect that the growing anti-Communist movement in America might have on the recruitment of able young men as scientists. The training of a scientist is long, arduous, and expensive. Would young men submit themselves to this kind of training when an accusation of disloyalty by an unknown might make them virtually unemployable overnight? As president-elect of the American Chemical Society he spoke often about the plight of the scientist threatened by overly zealous politicians.

Pauling was also one of a number of prominent scientists who were deeply disturbed by the threat of nuclear war. Prominent physicists such as Albert Einstein and Robert Oppenheimer had supported and assisted the development of nuclear weapons during World War II because they feared that Hitler's Germany might develop them first. They were appalled, however, when the bombs were first used on the defenseless and militarily insignificant cities of Hiroshima and Nagasaki in Japan. They feared that nuclear weapons would proliferate to many countries and that the eventual result would be a nuclear holocaust. A number of prominent scientists formed the Emergency Committee of Atomic Scientists to lobby for nuclear disarmament. Albert Einstein was chairman; Harold C. Urey was vice-chairman; Leo Szilard, professor of biophysics at the University of Chicago and a key member of the atomic bomb project, was a member; as was Harrison Brown of the University of Chicago's Institute for Nuclear Studies. Pauling had no personal involvement in the development of nuclear weapons, but he shared the concerns of the other scientists. He joined with Leo Szilard in speaking before audiences in Portland, Oregon, and Spokane, Washington, where they talked about the necessity for citizens to consider the proper use of atomic power. Pauling stated for the first time his belief that modern science had now made war untenable as a method of settling disputes among nations, and that more effective ways to keep the peace must be found.

Pauling was in great demand as a speaker for liberal political groups that wanted to capitalize on the prestige of the eminent scientists. He was eager to speak out on the issues, but Ava Helen felt that he often seemed ill informed when he ventured away from scientific topics. After a speech in 1945, she told him, "When you talk about the nature of war and the need for peace, you are not convincing, because you give the audience the impression that you are not sure about what you are saying."

Pauling had never felt comfortable expressing opinions unless they were based on solid information. When they were first married, Ava Helen remembered that when someone would ask him a question he would often refuse to answer, saying, "Sorry, I'm not an authority in that field." She took him aside and remonstrated, "If people would only speak when they are an authority we would not get anywhere."

Even at social occasions, Pauling had little patience for idle speculation. If he was going to speak convincingly on war and peace, he had to become an authority. So he set aside the time to read enough of the literature on the topics to make sure that his opinions were well documented. He formulated strong opinions that he could defend with confidence. He was in favor of multilateral disarmament. He believed in complete honesty and scientific accuracy in gathering data for use in diplomacy. He believed that the same moral codes should be applied to international relationships that are applied in ordinary social relationships of persons of integrity and goodwill. He supported world cooperation in raising total world living standards and in establishing the kind of world law that would prevent conflicts.

Pauling's method was one of moral exhortation. He believed that nations are not ordinarily moral, and that all nations need to mend their ways in that respect. Peace was not to be gained by recalling the past errors of other nations or by being self-righteous or selfishly nationalistic. What nations needed was to find a common ground for working toward bettering the whole world.

Perhaps because he thought his arguments so obvious, Pauling rarely questioned the motives of those who joined him in these struggles. When anyone gave him a petition, he read it carefully and signed it if he agreed with the views expressed. He didn't trouble himself to inquire into the ideology of the group that circulated the petition or to speculate about their ulterior motives. His name appeared on the petitions, letterheads, and public statements of dozens of organizations representing a range of pacifist and progressive viewpoints. Later on, when he sued the *National Review* for libeling him by calling him a Communist sympathizer, the *Review*'s attorneys listed many of these affiliations to defend their position.

Pauling particularly liked what Henry Wallace was saying in his Progressive party campaign and appeared publicly with Wallace at the Gilmore Stadium in Los Angeles in May 1947. The 18,000–seat stadium was jammed with 25,000 people, who heard Wallace speak of the need to turn our raw materials into tractors rather than munitions. Pauling was a vice-chairman of Wallace's group, the Progressive Citizens of America, and voted for Wallace in 1948. Ava Helen voted for Harry Truman.

Some of the policy issues involved scientific questions, particularly about the effects of atomic warfare. Pauling argued that the United States could not depend on maintaining world supremacy by an exclusive monopoly on atomic bombs. Other nations would eventually have the bomb and be able to use it. This was an important point at a time when Congress was busy considering legislation to keep the "atomic secret" from the Russians. Pauling and other prominent scientists kept pointing out that there was no such secret. Scientists around the world knew the fundamental scientific principles of the bomb, and many nations, including the Soviet Union, had the technological resources to develop nuclear weapons.

Pauling was not opposed to nuclear research or the peaceful use of atomic power. He was an advocate of nuclear power and thought it could be used to take salt out of sea water so the desert could bloom. Pauling's arguments were not limited to scientific points, however, but were fundamentally political. He advocated internationalism, world law under the United Nations, increased student exchange programs, worldwide relief and rehabilitation programs, and open sharing of scientific information. His views were not those of the majority of Americans during the early cold war years, but they were widely shared among the scientific and university community.

One reason for the liberalism of many scientists was the fact that they had international experiences and connections. The contrast between the simplistic anti-Communism of the United States and the more sophisticated political milieu in Europe was brought home to the Paulings in 1948 when they spent the winter and spring terms at Oxford University. Pauling had been awarded the prestigious George Eastman Lectureship and felt that this would be a good change of scene and a time for the family to get together. Linus Jr. remained behind at Pomona College, where he was a premedical student, but Linus and Ava Helen were accompanied by Peter, sixteen, Linda, fifteen, and Crellin, ten. Ava Helen said, "Some people complained about Oxford University being just for men, but I didn't have that experience. I got permission to go to any lectures I wanted and I had a very good time."[12]

The winter of 1948 was one of the most severe the English had experienced for many years. To the Pauling children, ice and snow were novelties. They found a furnished apartment near the campus and settled in. Ava Helen was free to attend lectures that interested her and felt comfortable on campus. The children found the schooling to be more individualized and stimulating, with less busywork. Achievement was encouraged, and all of the children did well. Crellin was soon first in his form in Latin and French and did well in algebra; these subjects were not taught to ten-year-olds in Pasadena. Peter's tutor, a Rhodes scholar, imparted his own zest for learning to the American adolescent. Peter was so taken with England that he returned for his college years, earned his doctorate in physics at the University of London, and remained there as a faculty member.

The political climate was like a breath of fresh air when compared to the conformist pressures of Southern California. Communist speakers were taken for granted in Hyde Park, together with the antivivisectionists, secularists, and advocates of free love, as well as the proponents of racial purity or the adherents of various religious sects. The English men and women with whom the Paulings talked could not understand why the United States with all its wealth and power, including sole possession of nuclear weapons, was so preoccupied with the fear of a handful of Communists in its midst. The Paulings had no real answer, except to suggest that the hysteria was compounded of a fear of possessing the atomic bomb and not knowing what to do with it, together with collective guilt for the bombings of Hiroshima and Nagasaki.

Pauling's hosts often made wry references to the American obsession with the Communist conspiracy. When Pauling spoke at the Natural Science Club at St. John's College, there was no reference to Pauling's part in protesting the witch hunt for subversives in the Los Angeles area. The menu, however, featured "un-American roast chicken." Pauling was not offended. He was at as much of a loss to explain what was going on in America as his English hosts. He was not accustomed to analyzing political or social forces or probing the depths of American national character. He was fundamentally a rationalist who believed that the source of error was faulty thinking and that the cure was a correct argument.

While travel was spiritually refreshing, it was physically exhausting and often required undue attention to tiresome details. Ava Helen made it possible for Linus to travel without becoming overly absorbed in routine matters. She helped him to remember people he had once met but half forgotten; made sure he dressed, ate, and rested properly; took messages and notes; and reminded him of schedules and appointments. On an exceptional occasion in late 1948 when he traveled alone to Paris, London, India, and Washington, D.C., Pauling returned home ill and exhausted. He never again traveled so far without his wife.

On the eve of leaving for his 1948 trip, Pauling had introduced the radical dean of Canterbury to a Pasadena audience. The dean, the Reverend Hewlett Johnson, was an early exponent of what was later called liberation theology. He was known for his book *The Socialist Sixth of the World,* in which he argued, "Communism has recovered the essential form of a real belief in God which organized Christianity, as it is now, has so largely lost."[13] His visit to the United States was quite controversial, with many anti-Communist groups demanding that he be denied a visa despite his high position in the Church of England. After Pauling introduced the dean, a local reporter sought to question him. He found Pauling back at his desk on December 14, 1948, facing a great stack of mail and unread student examination papers. Pauling was reluctant to take time for an interview, but the reporter persisted with a series of loaded and vaguely worded questions that implied that Pauling was an associate of Communists and subversives.

Pauling finally closed the interview by saying that although he personally disagreed with Dr. Hewlett Johnson, the dean of Canterbury, on a number of issues and agreed with him on others, that was not important. What were important was freedom of speech and the man's right to be heard. The next day's paper carried a long account of the interview, the import of which was that Linus Pauling, local professor, had refused to deny or affirm rumors that he was a "red sympathizer."

At this time, Pauling was at the peak of his career. His work was widely recognized in the Western scientific community. While he was becoming increasingly critical of American politics, there was as yet no assault on him from American authorities. Certainly there was no criticism of his science beyond the normal technical disagreements over specific issues from colleagues who respected his accomplishments. In late August 1951, however, a frontal attack on Pauling's scientific work was published in a major daily newspaper: *Pravda.*

It seems that two Russian chemists, Y. K. Syrkin and M. E. Datkins, had translated *The Nature of the Chemical Bond* into Russian and had added favorable comments of their own. The professional committee assigned to evaluate the volume made a negative report on its contents at a meeting of four hundred members of the Chemical Sciences Division of the Soviet Academy of Sciences. As a result of the committee's report, the entire body, with the exception of the two translators, censured the Russian chemists who had introduced Pauling's work to their fellow chemists, censured Pauling, and forbade the use of the volume in Soviet classrooms.

Pauling's work was characterized as decadent, pseudoscientific, and procapitalist. Apparently, the Russians equated his valence bond theory with philosophical idealism and feared that it would turn impressionable students away from dialectical materialism. Pauling was quite surprised by this attack, noting that the English Marxist philosopher J.B.S. Haldane had cited his theories as an example of dialectical materialism in science. He said that he doubted that anyone really understood both resonance theory and dialectical materialism well enough to say whether or not they have anything to do with each other. Pauling frequently remarked that he had read neither Marx nor Freud but he would like to if he ever found the time. Even in his undergraduate days, however, he had been "too busy" with the hard sciences to bother studying psychology, the social sciences, or the humanities.

Pauling felt that the attack on his science from the Russians was an example of what could happen when politicians dictate to scientists. He felt certain that the Soviet Union would suffer from not taking advantage of Western science. He thought the Russians were reacting to their isolation and their feelings of rejection by the West. In their attack on his work, they complained that Western scientists did not give enough attention to their own great chemist, Alexander Michailovich Butlerov.

While Pauling thought he understood the motivations behind the Russian attack, he was not prepared to overlook it. He was a master of polemical writing and probably wrote many of the press releases that were sent out by Caltech to publicize his talks and findings. On this occasion, Caltech sent out an eight-page press release that took the Russians to task for their criticisms. The press release was a masterful statement of Pauling's views, as well as a laudatory exposition of his accomplishments. The lead paragraph referred to "the famed and useful resonance theory of chemical bonds developed by Professor Linus Pauling." The press release then went on to explain how Dr. Pauling realized that troublesome molecules could be fitted neatly into the scheme of valence theory by postulating that they held two or three different valence-bond structures at the same time. While the neutral layman might easily sympathize with the Russians' feeling that it is inherently absurd to assume that something can be in two or more places at the same time, Pauling managed to make his theory seem eminently reasonable by capitalizing on the Russians' foolishness in attempting to resolve scientific questions through majority votes.

American scientists, also, had doubts about the adequacy of the valence bond theory, and were working on refinements, improvements, and alternative approaches, including, especially, molecular-orbital theory. These debates took place in the normal process of scientific interchange, however, and never in the form of an attack on Pauling's contributions, which were universally recognized as outstanding. In the United States, Pauling was attacked for his politics, not his science.

Pauling continued adamant in his belief in freedom of speech and in opposition to the attempt to force California teachers to sign loyalty oaths. He was in great demand as a speaker because of his scientific prominence and his oratorical skills, and he was more than willing to speak out both at local and more prominent gatherings, such as a February 1952 appearance in Carnegie Hall with the well-known progressive journalist I. F. Stone and others. This visibility caused a predictable storm of criticism from the McCarthyists and the right-wing press.

Pauling and Harold Orr, the president of the Los Angeles Federation of Teachers, were subpoenaed to appear before the California State Investigating Committee on Education because of their championship of teachers with alleged subversive political beliefs. Both men refused, under oath, to state whether or not they were Communists. Pauling, however, was not reluctant to state his political views publicly when he was not under oath.

"My own views are well known," he told the press.

> I am not a Communist. I have never been a Communist. I have never been involved with the Communist Party. I am a Rooseveltian Democrat. I believe that it is of the greatest importance that a citizen take political action in order that our nation not deteriorate. . . . A special loyalty oath would involve inquiry into political beliefs and would constitute violation of the fundamental principle of our Democracy.

In the 1950s, it took real courage to defend such views and Pauling rose to the challenge with vigor. When the regents of the University of Hawaii cancelled an invitation to him to dedicate their new science building, overruling their own president on grounds of Pauling's "leftist" political activities, he responded with a bitter countercampaign against the regents. He did not feel the least bit "unfit" to dedicate a science building, and he surmised that the men who acted in this way were surely doing their own jobs inadequately. In his view, the Islands deserved better regents than they had. They also deserved a better university than they had.

Pauling's stand was supported by many students and faculty at the University of Hawaii, as well as by many citizens who especially resented the accusation that he was "unfit" because he "hired a Jap during the war." Japanese-Americans in Hawaii had not been forced to leave their homes during the war years, and there were no incidents of disloyalty or sabotage from the large Japanese-American population in Hawaii. The Hawaiian section of Sigma Xi, a fraternity that included Pauling on its national executive committee, invited him to give a public address. He spoke in the still undedicated science building to an audience of unprecedented size, many of whom were disappointed to hear a rather technical lecture on the molecular structure of blood, metals, and intermetallic compounds rather than a political speech.

Pauling spent eleven days in the Islands, conducting his own investigation of the university and its policies as well as of chemical aspects of the Island's sugar industry. His visit was a tremendous boon to progressive political groups on the Island who were struggling against McCarthyism. Pauling's prestige as a scientist made him a valuable spokesman for such groups. Pauling had accepted this invitation in part because Linus Jr., by then a practicing psychiatrist, and his family were living there. As usual, work took precedence and he had very little time to play with his grandchildren.

Throughout this period, Pauling continued to be supported by members of his own profession, which showered him with professional awards and honors. He was the first recipient, on November 27, 1951, of the Gilbert N. Lewis Award, given to perpetuate the memory of the man who had been the direct precursor of Pauling's own work. Lewis had died, tragically, as a result of a laboratory accident. Dr. Wendell Latimore, professor of chemistry and former dean of the College of Chemistry, University of California, Berkeley, spoke at the award ceremony, describing Pauling as

> the leader in the broad fields of chemical philosophy. No other living chemist has contributed more to our understanding of the fundamental nature of the chemical bond and molecular structure. He has pictured for us many molecules, from the simplest to the most complicated silicates and proteins. More than that, he has elucidated the principles that have enabled us to understand why the atoms combine in their established patterns.

This view of Pauling's accomplishments was generally shared by his professional peers, who had given him virtually every honor and award that the profession had to offer. He also received numerous invitations to travel abroad to scientific conferences and felt a need to maintain contact with European chemists who were doing important work in the study of complex molecules. When Sir Robert Robinson, president of the Royal Society, invited Pauling to London to talk about his new protein research at a symposium, Pauling was eager to accept. He couldn't, however, be there on April 22, as requested, because he had to be in Philadelphia in his role as vice-president of the American Philosophical Society on that date. Sir Robert obligingly rescheduled the symposium for May 1, 1952.

Pauling was eager to attend, as he had missed several engagements in London before and the visit would give him an opportunity to confer with English scientists who were working in areas of great interest to him, including the study of the DNA molecule. He had to cancel some speaking engagements in France and disappoint the University of Toulouse, which was still waiting for him to pick up an honorary degree that it had awarded him the previous year. English scientists were eager for him to attend, and several other prominent institutions asked him to address their members. The trip had nothing to do with Pauling's political activities.

Pauling's regular passport had expired in January, so he applied for a routine five-year renewal. Inexplicably, the response was delayed. Finally, on April 21, Pauling was told that the State Department had decided that it was not in the best interest of the United States for him to go abroad. The State Department apparently was acting on the basis of information from the House Un-American Activities Committee, which mentioned him as a supporter of the "Communist peace effort."

Pauling did not stand on principle in this case, perhaps because the trip to London was so important to him. He went immediately to Washington and signed at least six loyalty oaths, as well as numerous affidavits saying that he was not, and had never been, a Communist. The only comment he could get from the three State Department officials with whom he spoke was that he had been found "not sufficiently anti-Communist." On April 28, a final refusal of his appeal was announced and Sir Robert had to be notified that Pauling couldn't come. Sir Robert responded with a most expostulatory letter to the *London Times.* The next day the refusal of a passport to Linus Pauling was world news.

Pauling was outraged at the insult to his personal reputation, as well as to the principle of freedom of travel. He was especially concerned because the passport ban interfered directly with his access to important scientific information. When his best efforts to pressure the government failed, Pauling returned to Caltech and decided to organize a conference on protein structures in Pasadena, to which he invited the most eminent English scientists with whom he would have met in London. Ironically, he obtained funds for this conference from the Office of Naval Research, through the American Institute of Biological Sciences, as well as from the National Foundation for Infantile Paralysis and the Rockefeller Foundation. The conference was held in September 1953 and was attended by Sir Lawrence Bragg and other important British scientists.

Pauling received support from many eminent persons in his continuing struggle for a passport. Albert Einstein wrote strongly commending him for the stand he was taking. Oregon senator Wayne Morse condemned the State Department's action from the floor of the U.S. Senate. Even former president Harry Truman remarked at a luncheon that he understood Pauling had been called before a congressional committee to give testimony about the nature of the red corpuscle. "I suggest," he remarked, "that Professor Pauling confine his investigations to the white corpuscle."

Pauling did not let the State Department's repression of his right to travel intimidate him from speaking out on the issues. In February 1953, he wrote to the president opposing the death sentence for Ethel and Julius Rosenberg. He protested the decision of the Atomic Energy Commission that denied J. Robert Oppenheimer access to classified information about atomic research, calling this the "worst case of national ingratitude I have ever known." In his commencement address at Reed College in 1954, where his daughter, Linda, was a member of the graduating class, he spoke out against politicians in general, who he felt were less objective and honest than the majority of scientists. Liberal colleges, such as Reed and Antioch, were eager to express their outrage at McCarthyism by inviting Pauling to speak.

From the State Department's point of view, the most disturbing criticism came from abroad. Pauling was invited to many international gatherings, and refusing him permission to attend was a slight to the inviters as well as to the invited. Pauling was honorary chairman of the International Congress of Biochemistry, which was scheduled to meet in Paris in July 1952. He applied for a passport once again for purposes of attending that meeting, and this time the State Department issued him a document valid for one trip only to this specific event. The State Department used this device again in April 1953, rather than deny Pauling permission to attend the ninth Triennial Solvang Congress in Brussels, where chemists were meeting to discuss the structure of proteins. He also received a limited passport to travel to medical schools in Germany on behalf of the Unitarian Service Committee, and to the International Congress of Pure and Applied Chemistry in July 1953 in Stockholm, where he was honorary chair of the physical chemistry section.

The Stockholm conference was the site of one of the most ironic episodes of this period. Pauling was attacked in front of fifteen hundred of the world's leading chemists by G. S. Zhdanov of the Russian delegation. Zhdanov criticized Pauling for his contributions to chemistry, as well as for his idealist and pacifistic leanings. The attack attracted great attention in the Stockholm press, which ran banner headlines about the "Russian-American Chemists' Duel," but the event was ignored by the

U.S. press. The European papers were partial to Pauling and critical of Zhdanov, and Pauling clearly won a propaganda victory for the Western world against the Communist bloc. The irony of this event was not lost on the Swedish public, and Swedish news commentators at this time predicted that Pauling, whose research on proteins was the talk of the Congress, would surely be the next recipient of the Nobel Prize in chemistry.

Pauling continued to receive limited passports for specific events, although some of them permitted considerable travel. In October 1953, he visited Israel for the ceremonies attendant to laying the cornerstone of the Weizmann Institute of Science. The State Department was apparently reluctant to affront the state of Israel and the American Jewish community by refusing a passport, but again, it was a temporary, one-visit document. A similar document enabled him to speak at the fifty-first session of the Indian Science Congress, a trip that enabled him to stop off in Greece and Israel in mid-December 1953. In February 1954, he stopped off in Honolulu on his return trip and again spoke to the American Chemical Society chapter in the still undedicated science building. He also gave an interview to a reporter in which he outlined a total development plan for the University of Hawaii. The university's president was not appreciative. "I have never met this man," he said, "and I don't care to hear about anything he has to say."

There is no doubt that Pauling's scientific work suffered from these political pressures, if only because they took so much time and energy. Of course, Pauling could have chosen to spend more time in the laboratory and less time traveling around the country and the world speaking on chemical and political topics. He had in the past been able to continue making significant discoveries while traveling and lecturing extensively, but this political repression was a new element and one that he found depressing.

The restraint of his right to travel may also have had a major effect by cutting him off from the data of English scientists that might have helped him to make a discovery that would have been a brilliant capstone for his career. Pauling was very close to being the first scientist to discover the structure of the DNA molecule (figure 15).

DNA (deoxyribonucleic acid) is found in all living things, encoding the genetic information needed for the transmission of inherited traits.

DNA is the "program" that guides the dynamic self-construction of an organism.

The structure of DNA, as would be determined by James Watson and Francis Crick, consists of two strands of a phosphoryl-deoxyribose polymer that are connected in a double helix by nitrogenous bases attached to the polymers. The four bases that occur in DNA—adenine, guanine, cytosine, and thymine—combine to form sixty-four possible three-base sequences or *codons*. According to Pauling's principle of aperiodic molecular coding of information, the data required to construct an organism are contained in the subtle one-dimensional patterns of codons. A *gene*,

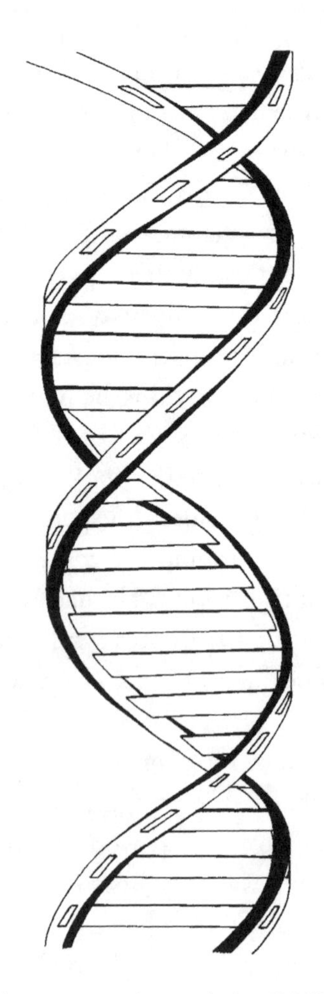

Figure 15. The double-helix structure that underlies DNA.

a segment of DNA that codes for the synthesis of a specific protein, generally consists of hundreds of different codons strung out in a row.

DNA replicates by separating into two single strands; each strand acts as a template for the construction of a new strand formed by the bonding of new bases to the original bases. Ultimately two new DNA molecules are formed, each half-old and half-new. This clever mechanism is responsible for the stable inheritance of traits from one generation to the next.

The race to discover the molecular structure of DNA has been one of the most publicized episodes in the history of science. The cast of characters and plot have been vividly, and controversially, portrayed by James D. Watson in a highly successful book, *The Double Helix,* which tells how he and Francis Crick rushed to find the structure and achieve scientific immortality before someone else beat them to it. Part of the competition was with Maurice Wilkins and Rosalind Franklin at Kings College, London, who had the funding, equipment, and expertise to produce the best X-ray diffraction pictures of crystals of DNA. Crick and Watson were working at Sir Lawrence Bragg's laboratory at Cambridge and under the British norms of respecting the turf of other scientists they really should not have been working on DNA at all. Bragg at one time had explicitly forbidden them to do so, but they continued in secret.

Much of the controversy has involved the ethics and politics of the relationships of Crick, Watson, Wilkins, and Franklin. The other silent, elusive character in the "race" was Linus Pauling. Watson and Crick were working very much within the framework of Pauling's theory of the chemical bond and relied on Pauling's book as a basic text. They also used Pauling's technique of model building, much to the disdain of Rosalind Franklin, who did not think of this as rigorous science. And they knew, through their contacts with Pauling's son, Peter, who had his desk in their laboratory at the time, that Pauling was interested in discovering the structure of DNA. Peter was very close to Watson, whom he had known in Pasadena. As Watson tells the story, he and Crick lived very much in fear that Pauling would beat them to the discovery that would win them the Nobel Prize.

Other participants and observers in the "race" have cast it in less vivid colors than Watson. The physical chemist Jerry Donohue, who worked in the same room with Watson and Crick at the time and supplied much of the chemical information that they needed to solve the structure, says that he did not perceive Watson as being in a race or working with great urgency. Donohue was a link between Pauling and Watson and Crick, in that he had worked for years with Pauling and was quite familiar with his methods and theories. Pauling, also, has denied being in a race with anyone, although he was interested in finding out the structure of DNA along with many other topics.

While Watson and Crick were working on the forbidden DNA model, Peter Pauling received a letter from his father saying that Linus was working on his own DNA model. Crick and Watson almost threw in the towel. Twenty-three-year-old Watson saw no reason not to go skiing in Switzerland during the Christmas holidays. They had no doubt that Pauling would soon have a paper and that it would be right. In Pasadena, on Christmas Day 1952, Pauling had a few guests to dinner who admired his model. He wrote to tell Peter that he had sent the paper describing a new structure for DNA on December 31 to the *Proceedings of the National Academy of Sciences* and that he had also sent a brief note to *Nature,* and another to Bragg. Would Peter like a copy of the note he sent to *Nature?* Peter said he would and they all waited.

On February 6, Peter's copy arrived and Crick and Watson both saw from Peter's face that it contained something extraordinary. When Watson looked at the Pauling paper he saw what Peter saw, that the structure was described as a triple helix and that the backbone was on the inside. Pauling's paper was not only wrong, it was all wrong. It didn't go anywhere; there was no adequate indication of how it replicated itself, how it provided for the carrying on of the species. Of course, Crick and Watson had also had the benefit of seeing Rosalind Franklin's X-ray crystallography, which showed that a three-stranded model was impossible.

Pauling and Corey's paper, "A Proposed Structure for the Nucleic Acids," which appeared in the *Proceedings of the National Academy of Sciences* in 1953, contained, in Watson's judgment, errors so fundamental that a student who had made them would not have been qualified to study at Caltech. Pauling later attributed his errors to the fact that he and Corey really weren't working very hard on the problem, as well as to errors and inadequacies in the published literature that provided the information that they used in constructing their model. While much attention has been focused on what Pauling might have learned from traveling to England, he could also have learned a great deal from people who were available in the United States such as Max Delbruck and Robert Sinsheimer on his own campus. Another invaluable informant who was in the United States at the time was Erwin Chargaff, who had published, in 1950, the epochal paper on nucleic acids that showed that in general the number of adenine units in each DNA molecule was equivalent to the number of thymine units, while the number of guanine units was equivalent to the number of cytosine units. This information was available to Pauling, but he forgot it. Watson and Crick remembered and were able to use it as the key to the structure.

In any event, Pauling and Corey did make it clear in their paper that they were publishing a tentative hypothesis and not a model that had been carefully evaluated. The fact that they published this poorly checked model at all gives some credence to the idea that they did feel that they were in a race with the English scientists, although it may have reflected nothing more than Pauling's tremendous self-confidence. On reading Watson and Crick's solution of the structure and seeing their experimental evidence, Pauling quickly recognized that they were correct and so stated.

When asked about the erroneous triple-helix model many years later, Pauling made the following comments:

> I got off on the wrong track, and Corey and I published the paper you mentioned. . . . These photographs showed that what Corey and I proposed wasn't right but . . . a different structure which involved a feature that I had been talking about for a number of years . . . complementariness between the two chains and I had said for several years that I thought the gene consisted of two parts which were mutually complementary.

He got this idea for two complementary structures, he said, "out of the antibody work."[14] It is true that Pauling's work in immunology made use of complementary molecular structures, similar to the ones involved in the double-helix model of DNA; and it is also true that Pauling pioneered the very idea of a nonperiodic helical structure. So it would not be surprising if the idea of a double-helical structure had at some point crossed Pauling's mind. But at any rate, when the crucial time came, he failed to develop this idea into a detailed model.

This controversy was later used by the influential science critic Bruno Latour to illustrate his "constructionist" theory of science. His book *Science in Action* includes a cartoon in which one scientist says, "The DNA molecule has the shape of a double helix." Another (presumably Pauling) replies, "Maybe it is a triple helix," while a third comments, "It is not a helix at all." Then they decide it would be "pretty" if it were a double helix, and it would explain Chargraff's work, so they agree to put that conclusion in the textbooks.

Whatever the merits of this caricature for Latour and his fellow cynics, it has nothing to do with Linus Pauling's approach. Pauling believed that scientific facts are established by reference to empirical evidence. He simply "got off on the wrong track" in his studies of DNA and published the paper with the erroneous model. When Crick and Watson's model came out, Ava Helen later remembered, he came home from work and told her simply that "the English got it right." He checked the facts, found his model to be faulty and theirs to be superior, and sent a letter to the journal retracting his conclusions. He also proposed some minor corrections in Watson and Crick's model, which were readily accepted. He took pride in Watson's acknowledgment, in his Nobel lecture, that "he and Crick were using my idea [of complementariness between the two chains], trying to make my idea definite."

Had the State Department permitted Pauling to visit England in May 1951, he almost certainly would have visited the laboratories at King's College where Wilkins and Franklin worked and probably would have seen their X-ray diffraction photographs. Many knowledgeable persons feel that had he seen these photographs and other data the English scientists had obtained, he would have had enough information to figure out the structure almost immediately. Pauling himself remarked, "I can't be sure what might have happened. . . . I knew Rosalind Franklin, and I might well have seen her and gotten an idea that would have put me on the right track." Pauling's associate, Robert Corey, did visit London at the time and saw the photographs, but he lacked Pauling's intuitive sense for molecular structures and did not come up with the flash of insight that Pauling might have had.

Of course, no one can ever do more than speculate on what might have happened had Pauling been allowed to carry out his plans to visit London. At the time, he was primarily interested in determining the molecular structures of proteins. He did not anticipate the special importance that the molecular structure of DNA would have for explaining the precise mechanisms by which genetic information is passed from generation to generation. Pauling read the literature on DNA and also had access to some X-ray diffraction photographs produced by Robert Corey in his own laboratories, though these photographs were not nearly as accurate as the ones he might have seen in London.

Ava Helen once asked him, "If that was such an important problem, why didn't you work harder at it?" Pauling's answer was "If I had worked harder, I wouldn't have needed to go to London to see Rosalind Franklin. I might well have discovered the double helix. I really wasn't paying much attention to the problem of the structure of nucleic acid."

Some people who were close to Pauling in this period say that much of the time he was depressed and preoccupied, an uncharacteristic mood for him. The political pressures were oppressive, and many political liberals were similarly depressed at the time. Now, not only were the political trends going against him, but others were outshining him in his own scientific bailiwick. The discovery of the double helix was based on theories and methods developed by Pauling along with others. It was a validation of the power of his theory of the chemical bond, and his ideas about biological structure, as represented in his model of the helical structure of polypeptides. His theories enabled Watson and Crick to elucidate the precise nature of the hydrogen bonds between the adenine-thymine and guanine-cytosine base pairs in the core of the helix that contain the specific information that makes up the genetic code. This opened up the way for a whole new field of biological science, including the potential for genetic engineering and creation of new species in the laboratory. It was a brilliant accomplishment, and Pauling had come close but had failed.

In the 1930s, Pauling had been the young scientist who made brilliant discoveries by building on the insights of others. Now younger scientists were claiming fame and fortune through the application of the principles that he had made so famous. While Pauling was quick to acknowledge that Watson and Crick were correct, he was not eager to accord them the eminence that their discovery merited. When Perutz asked Pauling to join him in nominating Wilkins, Watson, and Crick for a Nobel Prize, he refused, saying it was premature to give an award to Watson and Crick. Wilkins, he felt, didn't deserve one at all. The prize was awarded nevertheless, with Wilkins included for the quality of his pictures. Franklin had died prematurely.

6

Fallout, 1954–1960

On November 3, 1954, Linus Pauling was preparing to address the seminar Abnormal Hemoglobin Molecules in Relation to Disease at Cornell University in Ithaca, New York. The campus was electrified when the news came that their visiting lecturer had just been selected to receive the Nobel Prize in chemistry. Pauling received a standing ovation and reminded the participants that he had finished his classic work, *The Nature of the Chemical Bond,* at Cornell.

In making the award, the Swedish Nobel Committee did not single out any particular achievement such as the discovery of the alpha helix or the development of valence bond theory. Both of these achievements had been done in collaboration with or simultaneously with others. Instead, Pauling was given an unshared Nobel Prize in recognition of his "research into the nature of the chemical bond and its application to the structure of complex substances," a quite general statement that applies to the entire body of Pauling's published work in chemistry.

In making the award when they did, the Swedes asserted their neutrality by defying both the Soviet Stalinists and the American McCarthyists. Only one year had passed since the Soviets had attacked Pauling's chemistry at a highly publicized Stockholm conference, and the Swedes could not have made a more dramatic repudiation of their arguments.

Since Pauling would be invited to travel to Stockholm to pick up his award, their action was also a direct challenge to the United States government. Would Pauling be allowed to travel to Sweden to pick up the award?

Nazi Germany had interfered with the travel of Nobel Prize winners, and the Soviet Union would do so in later years, but the United States narrowly missed doing so. McCarthyism was coming to an end. On December 2, 1954, the U.S. Senate censured Senator Joe McCarthy by a 67 to 22 vote. On December 10, two weeks before the prize ceremony in Stockholm, Pauling received an unrestricted passport. The State Department responded to what its spokesman referred to as a "self-generated appeal," and the Pauling family made immediate plans to meet in Copenhagen and travel together to Stockholm. Peter and Linda flew in from Cambridge, while Linus Jr. and his wife took the polar flight from Los Angeles.

The American press took only perfunctory notice of Pauling's award, mentioning the fact that he had been "McCarthy's target." No American paper attempted a portrait of Pauling as a personality or of his family. In Sweden, however, the Nobel Laureates were given the kind of celebrity treatment usually reserved for stars of sports and cinema. They were accompanied everywhere by an attaché from the Swedish foreign office, who attempted to shield them from the attentions of the reporters who swarmed around them. The Swedish press, which Pauling could read fairly well, portrayed Linus as an example of what could be accomplished by a man of humble origins through hard work and intelligence. His lovely wife and four successful children were held up as a model family for the Swedish populace to emulate.

Although Sweden is in many ways a social democratic country, it retains the aristocratic traditions of a long monarchical history. The Paulings and the other Laureates—Max Born and Walter Bothe in physics, and John Enders, Frederick Robbins, and Thomas Weller, three Americans who had isolated the polio virus—were quite literally treated as royalty. The fact that the best known of the 1954 Laureates, Ernest Hemingway, could not attend because of an accident in Cuba made Pauling's appearance all the more newsworthy.

The presentation ceremonies began at four-thirty in the afternoon on

December 10. Each of the Laureates was ushered by his attaché into a small anteroom adjacent to the large auditorium to be introduced to the royal family. The Paulings were told that as Americans who were not used to the custom, they need not curtsy to the king. King Gustav Adolf VI was a kindly and approachable man, while his tense queen looked slightly distraught. They were accompanied by the widowed Princess Sibyla and her saucy, scarlet-clad daughter, the Princess Margaretha, and the dignified Prince Wilhelm.

Trumpets sounded as the king entered the main auditorium, followed by the royal entourage, then by the Laureates and their families. They sat in the front rows, where seats had been reserved in front of Sweden's social and diplomatic elite. Each of the Laureates' accomplishments was described by a specialist in his field. Professor Gunnar Haag's summary of Pauling's work virtually equated Pauling with the discipline of chemistry, stressing the crucial importance of understanding chemical bonds and the structures of substances in order to explain their properties.

A formal dinner with 760 guests in the magnificent Golden Room of Stockholm's showpiece Town Hall followed the ceremony. Hemingway would have been the dinner speaker had he been there since the recipient of the literary award is generally assumed to have the best command of language. In his absence, Linus spoke eloquently about the hydrogen bomb and the futility of war.

The remainder of the visit was a whirlwind of social events, speeches, and shopping trips, highlighted by an intimate dinner with the king and queen in their residence. The pattern continued for Linus and Ava Helen after they left Stockholm. They continued on an around-the-world trip with stops in Israel, India, and Japan. Linus's visit was a major event in each country. In Israel, he lectured at the Weizmann Institute, the Institute of Technology in Haifa, and the Hebrew University in Jerusalem. In India, he gave three lectures at the National Science Conference at Baroda and visited widely among Indian scientists.

In India, as in Sweden, the Paulings were the object of uncritical admiration. Their liberal political stand was appreciated by the Indians, who were critical of American and Soviet atmospheric bomb testing. Linus found that he and Jawaharlal Nehru were in general agreement about many points, including international relations, education, economics, and the status of women. Madame Pandit and Ava Helen were both members of the Women's International League for Peace and Freedom, a pacifist women's organization founded by Jane Addams. They too found much in common.

In Japan, Pauling's visit was sponsored by the Asahi Press, and hundreds of people had to be turned away from his public lectures in Tokyo and Kyoto. Pauling shared his knowledge of studies on the effects of radiation with the Japanese, who were especially bitterly opposed to nuclear testing. The Paulings spent three weeks in Japan, receiving uncritical adulation wherever they went. Then it was time to go back to the less admiring milieu of Southern California, with only a brief stop to see their grandsons in Honolulu while changing planes.

The California Institute of Technology is not a particularly liberal institution, and it is located in a community that was one of the strongholds of the conservative movements of the 1950s. Although Caltech always respected Pauling's academic freedom, he did not receive the warm support for his political activities that he found at liberal colleges such as Antioch, Oberlin, or Reed, where he was eagerly sought out as a speaker, or even at more established universities such as Cornell and Washington University in St. Louis, which have more of a tradition of political diversity. Pauling was, of course, a tremendous asset to Caltech in building its reputation as a center for basic research, and in helping to gain government and foundation support for that research. A faculty that includes a Nobel Prize winner would attract top-quality graduate students and young faculty. However, the administration found it relaxing to have its distinguished dissenting professor abroad for three months, and it hoped that the publicity elicited by his passport denial would die down.

Pauling remained outspoken, however, and relationships with the university administration were cool. One trustee resigned because of Pauling's position on nuclear fallout, and there is no way of knowing how much the institution suffered from the loss of local contributions caused by opposition to his politics. Officially, the institution took no notice of his political activities, although it did keep a copy of a sworn statement made by him to the effect that he was not and never had been a Communist on file in the public relations office.

Social pressures in the community were strong, and Pauling was especially hurt when someone whom he had thought of as an old friend passed him without speaking on a day when the *Los Angeles Times* reported on one of his statements. He and Ava Helen gravitated more and more toward the small and socially isolated liberal community in the Pasadena area. They became Unitarians, largely because the minister of the local church was a prominent community activist, and enjoyed the support of members of their new faith as well as of local Quakers, humanists, and members of the American Civil Liberties Union and of the Women's International League for Peace and Freedom, with which Ava Helen remained active.

The liberal groups were eager to draw upon Pauling's prestige, although there was some concern about taking so much of his time away from his science. Each year, the Pasadena branch of the Women's International League gave a pancake breakfast for fund-raising purposes at the Paulings' home. Pauling spent much of his time in the kitchen making pancakes. One of Ava Helen's European friends once admonished her for putting a man of his caliber to work at such a task, but he preferred that to making small talk with the guests.

There were always requests to speak, and Pauling accepted many of them. He spoke on electric power as a prime factor in the development of industrially backward nations at the Franklin Delano Roosevelt Club. His after-dinner talk at Caltech's Atheneum Club was on a more prosaic topic, "An Account of Our Trip around the World." The Citizens' Committee to Preserve American Freedom presented him with a Friendship Scroll decorated with the signatures of over three hundred persons from around the nation and the world. Pauling seemed eager to continue in the public eye, and the embattled liberal movements of the 1950s needed him badly.

During this phase of his life, Pauling continued to produce great volumes of scientific papers. While this research did not produce any discoveries on the order of the valence bond theory, the alpha helix, or the rules for mineral structure, it was nevertheless scientific work of very high quality. And, to his credit, Pauling was not content to rest on his laurels, producing minor modifications of his great breakthroughs of the past. Instead he continued to strike out in new directions. Certain of his research efforts were so innovative as to be far ahead of their time. An excellent example of this is his work on the molecular basis of brain function. Though the details of his papers in this area are generally no longer relevant to specialists in the field, his basic research program of "molecular psychobiology" was a very solid one.

Not until the 1980s did molecular psychology emerge as a respectable field of study. Yet it was only 1961 when Linus Pauling wrote the following words:

> During the last twenty years much progress has been made in the determination of the molecular structure of living organisms and the understanding of biological phenomena in terms of the structure of molecules and their interactions with one another. The progress that has been made in the field of molecular biology during this period has related in the main to somatic and genetic aspects of physiology, rather than to psychic. We may now have reached the time when a successful molecular attack on psychobiology, including the nature of encephalonic mechanisms, consciousness, memory, narcosis, sedation, and similar phenomena, can be initiated.[1]

Toward this end, Pauling outlined a novel theory of the molecular bases of anesthesia and a related theory of long-term memory. The story of Pauling's theory of anesthesia begins in 1959. One April morning, he was reading a paper by Pittsburgh crystallographers R. K. McMullan and G. A. Jeffery, reporting some work determining the structure of a crystalline hydrate of an alkylammonium salt. As Pauling wrote later:

> It is my memory that my thoughts during the next few seconds were the following: "This hydrate crystal decomposes (melts) at about 25 degrees Celsius. It contains dodecahedral chambers that might be occupied by xenon molecules, which would stabilize it enough to raise the decomposition temperature to above 37 degrees Celsius, the temperature of the human brain. Alkylammonium ions resemble substances normally present in the brain—amino acids, the side chains of protein molecules. Hydrate microcrystals involving these substances in the brain might form if the brain were cooled, or might interfere with the motion of ions or electrically charged protein side chains that normally contribute to the

> electric oscillations that are involved in consciousness and ephemeral memory, reducing their amplitude enough to cause loss of consciousness; or they might effect this result by interfering with some chemical reaction involved in supporting the electric oscillations, as by masking the active region of an enzyme molecule. The activity of anesthetic agents should be proportional to the polarizabilities of their molecules, which determine their effectiveness in stabilizing the hydrate crystals by the London electro-correlation intermolecular interactions; hence xenon should be more effective than argon, chloroform, more effective than methyl chloride, as observed. This is a molecular theory of anesthesia."[2]

Despite its frequent use of technical terminology, this passage gives us a rare glimpse at the inner workings of a great mind at a moment of creative invention.

Pauling's theory of anesthesia involves a large conceptual leap. On the one hand he observed the resemblance between alkylammonium ions and substances in the brain and inferred that, if alkylammonium ions could form hydrate microcrystals, so could the similar substances in the brain. On the other hand, he noted that the more effective anesthetic agents also were more effective at stabilizing hydrate crystals. These two observations led "naturally" to his hypothesis, that the stabilization of hydrate crystals is the process by which anesthetics affect the brain.

One point that stands out in this train of thought is the combination of conceptual rigor and free association. To a less inventive mind, it would never have occurred to relate the dodecahedral shapes of cavities of crystals with the difference between one anesthetic and another. On the other hand, to a less rigorous mind, this relationship, once conceived, would have remained a vague connection, not backed up by any testable details. To Pauling's mind, the facts of physical chemistry were instinctive, and the subtle patterns of crystal structures, quantum wave functions, and biological molecules were automatically perceived. All these tools were so natural and familiar as to be brought to bear instantaneously in the free-flowing movement from one idea to another.

Another point that stands out here is the easy transition between different levels of physical reality. In his classic work on the chemical bond Pauling bridged the two levels of physics and chemistry. He used quantum physics, for the first time, as an informal theoretical chemist's tool rather than as a mechanism for making detailed mathematical deductions. In his work on molecular biology, Pauling bridged the two levels of inorganic and organic chemistry. His hypothesis of aperiodicity identified an essential difference between life and nonlife and at the same time clarified the numerous parallels. And in this speculative theory of anesthesia, he bridges the two levels of molecular chemistry and neuroscience. In the heat of discovery, the two different contexts are exactly the same—the important quality being, not the level of the system, nor the terminology used to describe it, but the specific geometrical and physicochemical properties involved.

The brain is, in essence, an electrochemical system. The most important cell in the brain is the neuron, and what the neuron does is store and transmit electricity. Essentially, a neuron converts voltage into frequency: the more charge one feeds into the neuron, the faster it puts out its short bursts of charge. It sends these bursts of charge to other neurons, but the connection is not direct: in most cases, there are intervening chemicals called *neurotransmitters.* When a neuron fires, it causes certain neurotransmitters to transmit this charge to other neurons. These neurons, if their total charge is sufficient, may fire in turn and set off other neurons, et cetera. Eventually some of these indirectly triggered neurons may feed back to the original neuron, setting it off yet again, and starting the whole cycle from the beginning. Thoughts, memories, and feelings may thus be envisioned as reverberating circuits of electricity in the brain. But all of these circuits are dependent on the underlying chemistry, and this is where Pauling felt he had a contribution to make.

According to Pauling's colleague J. F. Catchpool, anesthesia is naturally defined as "a reversible depression of the level of consciousness." Pauling developed his initial intuition regarding anesthesia by focusing on the water contained within the neurons and in the regions between the neurons. Hydrates forming in this water, he proposed, may capture some of the electrically charged side chains of neuronal proteins and interfere with the movement of charge through the neurons, thus cutting off the reverberating circuits to the extent that consciousness is lost. This idea, based on the correlation between anesthetic potency and ability to form hydrate crystals, was in many ways an improvement on the traditional theory of anesthesia, proposed by pharmacologists Hans Meyer and C. H. Overton at the turn of the century: that anesthesia occurs when any substance attains the right concentration in the fatty portion of the cell.

And, while thinking about anesthesia, Pauling also had his eye on a larger problem: the nature of memory. He was aware of the "two-store" theory of memory, in which long-term, unconscious memory is distinguished from short-term, conscious memory, and of the relation between these types of memory and anesthesia. Patients entering or emerging from anesthesia pass through a stage during which they have short-term memory and can recall previously stored long-term memories but cannot actually form any new long-term memories. So it seems that anesthetics, in subanesthetic doses, act specifically to inhibit long-term memory. Experiments done on chicks, by Pauling's colleague Arthur Cherkin, verified this phenomenon.

Cherkin proposed the additional hypothesis that anesthetic molecules, in small quantities, might actually enhance the production of long-term memories. This train of thought is very much in line with current trends in cognitive neuroscience: today hundreds of researchers are seeking the molecular substrate of long-term memory formation. But at the time Pauling's call for research in this area went largely unheeded, and Pauling himself left the work unfinished, moving on to other things.

Cherkin, however, relates an amusing story regarding the work on anesthesia. Pauling had one group of researchers investigating anesthesia in brine shrimp and another studying anesthesia in goldfish. Both groups were experimenting with a variety of different anesthetics, in each case measuring a quantity called the *anesthetic partial pressure.* When the goldfishers compared their values with those of the brine shrimpers, they found close agreement for cyclopropane, ether, halothane, and chloroform. But the values for n-pentane were way off, casting doubt on the viability of the theory. Careful rechecking of the work revealed no errors but did show that, because of certain technical factors, the value for the solubility of hydrocarbons in water entered into the calculations for n-pentane, and not for the other anesthetics.

But then, at that very moment, American soil scientist Clayton McAuliffe published new figures for the solubility of hydrocarbons in water, which forced the brine-shrimpers to revise their figures upward. All the figures were now in order; the theory looked hopeful once again. Pauling remarked sarcastically, "Hmm, perhaps you should use goldfish to measure solubilities." For, on the basis of his theory of anesthesia, Pauling could have predicted that the old value for the solubility of hydrocarbon was wrong!

Then he made a familiar comment: "You know, you never want to let a 'fact' stand in the way of a good theory." Coming from a man who had originated so many good theories and had also discovered many important empirical facts, this statement has a great deal of meaning. It reveals a deep faith in the logical structure of the world. Good theories are formulated on the basis of informed intuition, on a combination of deductive, inductive, and analogical reasoning. Pauling's theory-formation process was guided, in particular, by his comprehensive knowledge of physics and chemistry and his intuitive understanding of quantum-theoretic and crystallographic structure. When it came right down to it, his faith in these abstract structures exceeded his faith in the numerical results of any particular experiment: An experiment could always be in error, but there was no mistaking the logical order of the world.

As they moved into their sixties, Linus and Ava Helen naturally began to think of retirement and actually went so far as to purchase a 160-acre tract in the Big Sur country along the majestic California coastline. They came across the small parcel of land by accident while driving through the countryside adjacent to the Hearst estate and bought it on impulse from an elderly resident who wanted to move into town because of failing health but was eager to keep his property separate from the Hearst domain. The small cabin on the property, without electricity or telephone, was one of only four homes along the ocean for the fifty miles between San Simeon and Lucia.

The Ranch, as they called it, was a delightful place to escape for long weekends, and their children and grandchildren loved to visit there. As pleasant as it was for these retreats, it was not enough to provide a fulfilling life for a man who loved controversy and appearances before audiences. The Paulings' sons and daughters predicted that some cataclysmic

Linus Pauling at one year of age. *Used by permission of Linda Pauling Kamb.*

Linus Pauling in the 1930s, when doing research on the chemical bond. *Courtesy of the Archives, California Institute of Technology.*

Ava Helen Miller Pauling, Zurich, ca. 1925. *Used by permission of AIP Emilio Segrè Visual Archives, Goudsmit Collections.*

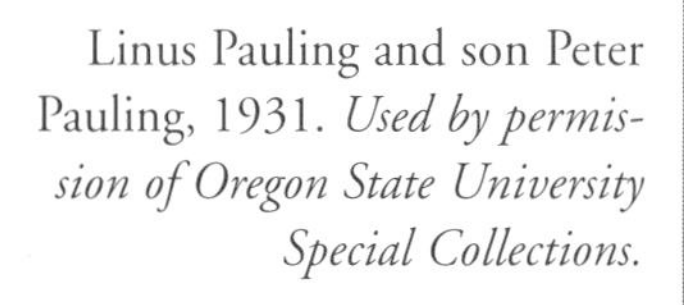

Linus Pauling and son Peter Pauling, 1931. *Used by permission of Oregon State University Special Collections.*

Ava Helen Pauling, Linus Pauling, Jr., Linda Pauling, and Peter Pauling, 1934. *Used by permission of Oregon State University Special Collections.*

Linus Pauling at work in his laboratory at Caltech, 1940. *Used by permission of Oregon State University Special Collections.*

Linus Pauling, formal portrait. *Courtesy of the Archives, California Institute of Technology.*

Linus Pauling with model of molecular structure. *Courtesy of the Archives, California Institute of Technology.*

Linus and Ava Helen Pauling, passport photograph, 1960. *Used by permission of Oregon State University Special Collections.*

Linus Pauling reading *No More War! Photo by Kendall Brown, courtesy of the Archives, California Institute of Technology.*

(19)

Only mass protest can achieve the goal.

I believe in non-violence. But the Establishment believes in violence, in force — in MACE, NAPALM, Police power, aerial bombing, nuclear weapons, war.

So long as the selfishness of the Establishment remains determinative, our hope that the coming revolution will be non-violent has little basis in reality.

The End

Excerpt from a speech given in Los Angeles in 1968. *Used by permission of Oregon State University Special Collections.*

Linus Pauling in the classroom. *Used by permission of Oregon State University Special Collections.*

Linus Pauling on his seventy-fifth birthday between his son Crellin (*left*) and his grandson Barclay J. Kamb, Jr. *Front L–R:* Anthony Kamb, Ava Helen Pauling, Linus Kamb, Linda Pauling Kamb, Linus Pauling, Jr. *Courtesy of Archives, California Institute of Technology.*

Arthur and Laurelee Robinson. *Used by permission of Arthur Robinson.*

Linus Pauling with his research notebooks at his home, 1987. *Used by permission of Oregon State University Special Collections.*

Linus and Ava Helen Pauling at the Women's International League for Peace and Freedom meeting, Santa Cruz, California, 1979. *Photograph by Leonard A. Greenberg, used by permission of Oregon State University Special Collections.*

world event would surely preoccupy their parents and that they would seldom visit the Ranch.

The world event that increasingly drew Pauling's attention was nuclear bomb testing in the atmosphere. It was an issue of considerable scientific controversy. Biologists and geneticists were critical of testing on the grounds that the nuclear fallout would cause cancer and genetic defects, while the physicists were bitterly divided on the issue. The Atomic Energy Commission (AEC) minimized the dangers and questioned the motives of the scientists who opposed testing. The promilitary physicist Edward Teller insisted that an illuminated watch dial could cause more harm than a nuclear test. In any event, smog could protect American cities from fallout.

Pauling joined many eminent scientists in opposing testing. George Leroy, associate dean of biological sciences at the University of Chicago, charged that the AEC was withholding information from the public. Warren Weaver, a former president of the American Association for the Advancement of Science, testified in Congress that H-bomb testing would be the cause of physical defects in six thousand babies in one year and even more in years to come. His testimony was not, however, covered in the newspapers. One of the most effective attempts to document and publicize the effects of fallout was that of Professor Barry Commoner at Washington University, who collected baby teeth from children all over the country and demonstrated the presence of radioactive strontium 90, which did not occur naturally but could come only from nuclear testing.

Pauling visited Washington University in St. Louis in May 1957 to give the lecture "Science in the Modern World." More than a thousand students and professors heard him at a morning convocation that honored members of academic fraternities. He spoke eloquently about the dangers of radioactive fallout and argued that the advances of science as applied to war had made war obsolete. The audience was unusually responsive and wanted to do more than just listen and agree. A group met in the afternoon, and by evening copies of a petition of protest, written by Pauling with the help of the distinguished scientists Barry Commoner and E. U. Condon, were ready for the mail.

The text of the petition was as follows:

We, the scientists whose names are signed below, urge that an international agreement to stop the testing of nuclear bombs be made now.

Each nuclear bomb test spreads an added burden of radioactive elements over every part of the world. Each added amount of radiation causes damage to the health of human beings all over the world, and causes damage to the pool of human germ plasma such as to lead to an increase in the number of seriously defective children that will be born in future generations.

So long as these weapons are in the hands of only three powers an agreement for their control is feasible. If testing continues, and the possession of these weapons spreads to additional governments, the danger of outbreak of cataclysmic nuclear war through the reckless action of some irresponsible national leader will be greatly increased.

An international agreement to stop the testing of nuclear bombs now should serve as a first step toward a more general disarmament and the ultimate effective abolition of nuclear weapons, averting the possibility of a nuclear war that would be a catastrophe to all humanity.

We have in common with our fellow men a deep concern for the welfare of all human beings. As scientists we have knowledge of the dangers involved and therefore a special responsibility to make these dangers known. We deem it imperative that immediate action be taken to effect an international agreement to stop the testing of all nuclear weapons.

The petition was signed by Linus Pauling and twenty-six other "first signers" and circulated to scientists throughout the country. Volunteers from Washington University ran the mimeograph machines and sent out the first mailings from a list Pauling supplied. He knew U.S. scientists well enough to know who would be responsive. Within less than three weeks, two thousand signatures had been obtained, including those of the Nobel Laureate H. J. Muller, who had received the prize for discovering that radiation causes mutations, and L. J. Snyder, an eminent geneticist. Snyder signed as an individual, but it did not escape notice that he was also president of the American Association of Science.

Other important scientists disagreed with the petition, not necessarily because they denied that radiation caused damage but because they considered the damage minor enough to be acceptable in light of the need to protect the security of the free world. After all, the deaths that resulted from nuclear testing were fewer than those from automobile accidents. One professor even suggested that the testing might be useful in slowing the world's population increase. At its core, the argument was more political than scientific. All scientists agreed that the testing could cause damage, but no one could say precisely how much. The pro-testing scientists used the minimum estimates and said the risk was justified for military reasons. The antitesting scientists warned that the maximum estimates might be correct, and in any event escalation of the arms race was not the way to world peace.

Pauling's contributions to the debate were political and educational. He was not personally involved in research on the effects of radiation, although he did read the literature and used it to buttress his arguments. His prestige as a Nobel Prize winner, combined with his outstanding skills as a speaker, made him a focal point of the movement. It also attracted the attention of Senator James O. Eastland's Internal Security Committee, which sent Pauling a subpoena to appear before it on June 18, 1957. The committee was concerned about possible Communist influences, or at least Communist "inspiration," in the petition movement.

Pauling did not get his chance to appear before the committee on this occasion, although he expressed eagerness to do so. He had a speaking engagement at the Pasteur Institute in Paris on that date and requested a postponement. The committee, responding to the deluge of protest that followed the announcement of the subpoena in the press, postponed the hearing indefinitely.

As Pauling traveled throughout Europe that summer, including a trip to Moscow for a biochemistry conference, he found that foreign scientists strongly supported his stand against testing. He decided to initiate an international petition drive from his home in Pasadena. U.S. critics had claimed that he had already enlisted all the scientists who supported his position, and he was eager to demonstrate that eminent scientists around the world were overwhelmingly behind him.

This turned out to be easy to do. With the help of Ava Helen and a part-time secretary, Pauling mailed out letters to scientists around the world, and petitions flooded in. Some respondents sent only their own signature, but others had the petitions translated into their own native languages and distributed them widely. The average recipient sent back fifteen signatures. The total cost of the international petition campaign was $250 for stamps, printing, and clerical help. They had spent $600 on the U.S. petitions. While the petitions were coming in, Pauling kept the issue in the news by speaking at a mass rally in a football stadium in Van Nuys, California; at an international colloquium in New York; and at other events.

On January 15, 1958, Pauling presented the names of 9,235 petitioners from forty-four countries to Dag Hammarskjold, secretary general of the United Nations. The final count was 11,021 names from forty-nine countries. Thirty-six of the signatories were Nobel Prize Laureates, 35 were members of the Royal Society of London, 216 were members of the Academy of Sciences of the Soviet Union. Americans accounted for 2,705 names, including two of special importance to Linus and Ava Helen: Peter Pauling in England and Linus Pauling, Jr., in Hawaii.

This overwhelming support from the world scientific community did not result in immediate acceptance at home. The *New York Daily News* argued that the petition "peddles the familiar and exploded Communist line about fallout from those tests endangering future generations." Fulton Lewis, Jr., a conservative columnist, raised doubts about who was really behind the petition movement. He claimed that experts had told him that assembling such a petition would cost an average of $10 per signature and publicly wondered where the $100,000 came from. However, the opposition was beginning to crack. The Atomic Energy Commission was forced to admit that a small atom bomb they had set off in Nevada in March 1958, in an attempt to prove that a similar test done by the Russians could not be detected, was actually recorded by seismic stations as far away as Alaska. During a debate with Pauling over San Francisco radio station KQED, Edward Teller agreed that there was no such thing as a "clean bomb," although he still defended testing as necessary to defend freedom.

This concession did not stop Teller from giving an interview to *Life* magazine that *Life* announced on its cover as a refutation of the nine thousand scientists who had spoken out against the test ban. Pauling read the article and concluded that Teller had made no such refutation. He attempted to get space in *Life* for a reply but was refused. Similar refusals came from *Look, Saturday Evening Post, Ladies Home Journal,* and *Reader's Digest.* An editor at *Harper's* also refused to publish Pauling, arguing that both Teller and Pauling were at fault for issuing "oracular and highly debatable pronouncements on international affairs."

Frustrated by the magazine rejections, Pauling decided to write his first popular book. *No More War!* was dictated furiously over two consecutive four-day weekends, and the results were rapidly transcribed by three typists. The book was an elaboration of arguments he had made in speeches about the dangers of nuclear fallout, combined with a general argument that nations should stop preparing for nuclear war and settle their differences peacefully.

Although the book includes a brief scientific explanation of the nature of nuclear weapons, the only topic that is treated in real depth is the danger of nuclear fallout. This was a controversial issue, with Edward Teller and scientists from the Atomic Energy Commission claiming that the danger was small and acceptable, while Pauling and his supporters said that it was unacceptable even though they conceded that it was small in comparison to many other risks.

Pauling told his readers that the AEC claimed that the harmful effects of nuclear fallout were less than those of cosmic rays or natural radiation, and that the risk of having a defective child would be increased more by moving from a wood house to a stone house, or by moving from San Francisco or New Orleans to Denver or another high-altitude city, than by being exposed to fallout from bomb tests. They argued that an airplane pilot incurred a greater risk of having a defective child from the radioactive dials of the instruments in his cabin than from fallout.

After listing these and several other claims made by the AEC, Pauling went on surprisingly to concede, "All of the above statements are true." But they did not "tell the whole story." He argued:

> I believe that one goal that human beings strive for is that of decreasing the amount of human suffering in the world, and that it is worth while to ask how many individual defective children will be born in the world as a result of the testing of nuclear weapons.

Of course, any increase in human suffering is unacceptable if it is not balanced against an equivalent gain. The difficult issue is balancing the costs of technology against the benefits. Pauling acknowledged, for example, that the radiation exposure from medical and dental X-rays was much greater than that from fallout, but he recognized that the medical gains often justified the risk. Pauling's opponents, especially Edward Teller, argued that the benefits to national security justified the risks of fallout. Pauling rejected this argument largely on moral grounds:

> I believe in morality, in justice, in humanitarianism. We must recognize now that the power to destroy the world by the use of nuclear weapons is a power that cannot be used—we cannot accept the idea of such monstrous immorality.
>
> The time has now come for morality to take its proper place in the conduct of world affairs; the time has now come for the nations of the world to submit to the just regulation of their conduct by international law.

No More War! did address some scientific controversies. One was the question of whether there is a *threshold effect* in radiation damage such that exposure below a certain threshold causes no damage. Another was whether the fact that no higher rates of leukemia or birth defects could be documented in high-altitude locations such as Denver or Tibet suggested that the danger from fallout was negligible. Pauling addressed these scientific issues directly and in many cases conceded that there was simply no certain knowledge. There was, however, some imbalance in his treatment of these scientific uncertainties. He insisted that his opponents "prove" their points and rejected their arguments when they failed to do so. Yet he based many of his own conclusions on his own best judgments or estimates, even when they could not be conclusively "proved."

The more fundamental difference between Pauling and his opponents, however, was their judgment on international affairs. Teller and his conservative supporters believed that the security of the United States depended on containing Communism, and that the best way to do this was to maintain a strong nuclear arsenal. Pauling and his liberal supporters believed that the arms race would lead to the spread of nuclear weapons and an increased likelihood of nuclear war.

This was a crucial policy issue, but Pauling did not address it in detail. Instead, he offered rhetorical conclusions:

> The time has now come for war to be abandoned, for diplomacy to move out of the nineteenth century into the real world of the twentieth century, a world in which war and the threat of war no longer have a rightful place as the instrument of national policy. We must move towards a world governed by justice, by international law, and not by force.

Pauling did not offer any substantial arguments to assuage the doubts of those who doubted that this appealing vision was more than a seductive fantasy. Nor did he discuss in any depth the risks that attempting it might have entailed. He insisted, "Any national policy that depends upon ever greater power of destruction . . . followed by two great nations, is sure to lead to catastrophe for the world." This seemed self-evident to him at the time, as it did to many Americans with similar political views. From our vantage point in the post–cold war era, however, it must be conceded that the policy of using the nuclear threat to contain Communism until it collapsed of its internal contradictions was ultimately successful.

Of course, there is no reason why Linus Pauling, a physical chemist, should have been expected to resolve these social and political issues singlehandedly. Appropriately, Pauling ended *No More War!* with a proposal to establish a World Peace Research Organization, within the structure of the United Nations. This organization would be staffed by specialists in the social as well as the physical sciences and would attack the problems "in the way that other problems are attacked in the modern world—by research, carried out by people who think about the problems year after year."[3]

Toward the end of March, Pauling received a letter from a young editor at Dodd, Mead and Company asking whether he would be interested in writing a book stating his views on foreign affairs. A contract was quickly signed, and the editor was no doubt surprised to receive a manuscript by return mail.

The Paulings continued to use a number of approaches to publicize their opposition to testing. Linus joined seventeen other persons as plaintiffs in suits against the governments of the Soviet Union and the United States charging that those nations had violated rights guaranteed them by the charter of the United Nations. They could not sue Great Britain, because permission from the queen was not granted. In May

1958, Pauling and Nobel prize-winning biochemist Albert Szent-Gyorgyi of Woods Hole, Massachusetts, asked the U.S. National Academy of Sciences and their counterparts in England and the Soviet Union to establish a scientific world parliament that would make recommendations to assist nations in solving great international problems. While giving a graduation address at Antioch College, he heard that he had been made a member of the Soviet Academy of Science, together with a number of other world figures in science. Several newspapers and magazines attributed this new honor to his prominence in the antitesting movement, although that movement was as much opposed to Soviet as American testing. Pauling interpreted it as a sign that the Soviets were learning not to mix science and politics. He had triumphed over the Soviet scientists who had attempted to ridicule his science, and the tide seemed to be turning against his domestic opponents as well.

To many observers, Pauling's mood seemed ebulliently positive throughout the period following the Nobel award. Things seemed to be going well both in the Soviet Union, where scientists were winning increased freedom from political control, and in the United States, where more and more people were speaking out against anti-Communism and against nuclear testing. His life was largely absorbed in public affairs at this time, and he found it exciting and personally rewarding as he traveled around the world speaking to admiring crowds and jousting with his opponents.

The fly in the ointment was the situation at Caltech. Under pressure from the president and trustees of California Institute of Technology, he had resigned the chairmanship of the Division of Chemistry and Chemical Engineering at Caltech, a position he had held for twenty-two years, in June 1958. He could not be fired from his tenured position, but he was denied raises when others got them and suffered a cut in pay when he was no longer division chairman. He was also forced to give up laboratory space on the grounds that others needed it more.[4] Although he continued as director of the Gates and Crellin Laboratories, as well as professor of chemistry, chemical research was no longer the primary focus of his life.

In January 1959, Pauling engaged in a controversy with a scientist who had done some experiments with mice that she claimed indicated that ingested strontium 90 was harmless. She published her findings in *Science,* and Pauling replied with a note to the *Proceedings of the National Academy of Science* criticizing her for using far too small a sample to detect the effects of radiation on the genetic system. His reply was coauthored with a young Caltech assistant professor of geology, Barclay Kamb, who later became Pauling's son-in-law.

Pauling first became interested in Barclay Kamb when the young man tried to enroll at Caltech at the age of fifteen. Pauling was impressed with his potential at that time and tried to get him admitted but failed. Kamb was admitted the following year, and Pauling followed his career with interest. Linda Pauling had heard her parents speak of Barclay so often that she was resistant to meeting him, and he was reluctant to become involved with the daughter of his favorite professor. Nonetheless, they fell in love, married, and continued to live in Pasadena in close association with Linus and Ava Helen.

The Paulings continued to travel incessantly. In June 1959, they were abroad for eight months visiting England, Germany, Norway, Sweden, Japan, and French Equatorial Africa. He was no longer engaged in research that would be of great interest to professional chemists who had read about his valence bond theory and other contributions to chemistry in their textbooks. He did give lectures, however, primarily on the topic of the application of scientific research to medicine. He often spoke at hospitals or addressed large audiences with varying levels of scientific expertise at universities.

They also attended many political rallies and spoke on nuclear testing and world peace. Ava Helen was a delegate, and Linus a speaker, at the Women's International League's international conference that met in Stockholm that year. They were pleased with the ability of women from all over the world to meet in the pursuit of the peaceful resolution of international disputes.

The Paulings then made the long journey into the West African jungle especially to meet Albert Schweitzer at Lambarene. Schweitzer was eagerly waiting their coming. In April 1957, he had taken time off from his medical and administrative duties at the hospital to prepare himself to speak on the subject of nuclear testing. He studied the mechanisms of the creation of radioactive elements in a nuclear explosion, the injection of radioactive dust created in atomic explosions into the upper atmosphere, the patterns of fallout, the routes of assimilation of radioactive isotopes into living tissue, the modes of genetic and somatic damage by high-energy radiation, and the half-lives of the radioactive elements. He studied the available information about the probable effects of the explosion of megaton bombs in heavily populated areas in a nuclear war, scarce studies only a few military men had bothered to make.

On April 24, 1959, Schweitzer dramatically added his name to the list of world leaders who made a statement against the continued testing of nuclear weapons. His speech "Peace or Atomic War" was broadcast from Radio Oslo in Norway. Radio listeners all over the world heard his message in rebroadcasts, except in the United States, Great Britain, and the Soviet Union.

The Paulings, like all guests who went to Lambarene, approached the hospital in the jungle in a canoe. They climbed up the muddy bank to grasp the genial doctor's outstretched hand. The plain little guest room to which they were taken by a growing entourage was cool and clean.

Inside the low buildings were the staff quarters that housed other foreign visitors, who, like Schweitzer, had come to wash the feet and bind the wounds of the Africans. A colony of lepers is housed nearby. Some of the stronger lepers, who found walking on their infected feet difficult, got great satisfaction from paddling the canoes that took guests to and from Lambarene.

While the Paulings were impressed with the services that Dr. Schweitzer and his colleagues were providing, they did not fit comfortably into the paternalistic ethos of the place. Others had criticized Schweitzer for not training the natives as nurses or "barefoot doctors," and the Paulings noted an authoritarian atmosphere even at the dinner table. Delicacies were placed in small dishes near Schweitzer's plate that were for him only. Ava Helen, whom he liked very much, was seated beside him and unknowingly helped herself from dishes intended only for Schweitzer. An awkward silence among the staff informed her that she had been gauche, although Schweitzer urged her to take more from his small dishes.

In a less immediate way, it was Linus who played the role of the person who goes to a friend's house for dinner and hires the cook. Pauling made a strong impression on a young English physician, Frank Catchpool, who was volunteering for Schweitzer while recovering from an unhappy love affair. Not long afterward Catchpool began to do research with Pauling. He remained close to Pauling and was the principal character, other than Pauling himself, on a public television special devoted to Pauling. He went on to practice medicine in Sausalito, California.

The Paulings continued their trip around the world, stopping in November for the opening of the Australian and New Zealand Conference on Peace and Disarmament. The conference was full of bitter arguments among pacifists, socialists, trade unionists, and other iconoclastic individuals. They spent three weeks in Australia, during which time Ava Helen gave thirty-four speeches on peace and Linus spoke at the major universities. His speeches focused on the dangers of nuclear testing, although he felt that Australia would be the least affected by nuclear fallout. He anticipated a gradual convergence between the social systems of Communist and capitalist countries and regretted that more Communists had not attended the peace conference.

In June, in Hiroshima they joined twenty-three other distinguished world citizens who were invited by Kaoru Yasui, dean of the law faculty at Hosei University and director-general of the Japanese Council against A and H Bombs, to participate in a pilgrimage to Hiroshima. The entire city paused for a moment of silent prayer on the fourteenth anniversary of the American bombing of their city, and the Paulings were part of a crowd of 340,000 people. Two Hiroshima maidens, disfigured by the bomb when they were children, tolled the memorial bell in the Peace Plaza.

On December 5, 1959, the Paulings returned to Pasadena and their usual activities. Linus spoke to the FDR Club about his visit with Schweitzer. Ava Helen took the pulpit one Sunday morning to tell the members of the First Unitarian Church in Los Angeles about the Peace Congress in Australia. The most traumatic event of this period was an hour on the morning of Sunday, January 31, when thousands of radio listeners thought that Linus was dead.

The Paulings had gone to the Ranch that weekend for one of their all too infrequent visits. Pauling left the cabin on Saturday morning and went out to check the fences on the ranch. Sometime late in the afternoon he had to admit to himself that he could not get back to the cabin. While looking for a possible route around a cliff to pipe water into his house from Salmon Creek, he had clawed and scrambled up a steep deer path only to find himself at a dead end on a narrow rocky ledge on the cliff. One hundred feet below, down an 80-degree incline, were the ocean and jagged rocks. As he turned to start back, the greasy blue shale slipped and slid beneath him, so he scrambled back to the ledge. The thing to do at a time like this, he knew, was to stay put and wait to be found. If he fell, he might lie seriously injured for hours or days or be killed outright.

At six o'clock that same Saturday evening, Ava Helen went to the ranger station to ask for help. Her husband was a cautious man who had left home thinly clad. The sun had set. If he were not in trouble, she knew he would already have returned to the cabin. The ranger immediately sought help: bloodhounds, the county sheriff, youthful volunteers, and two helicopter pilots from nearby Fort Ord. Soon after dark, Pauling heard voices and saw flashlights. He thought his speedy rescue almost too good to be true and shouted loudly. The noisy surf below drowned out his voice. In a few minutes, he was again alone in the silent foggy chill darkness.

Pauling scooped out a hollow on the narrow ledge and covered himself with a big map he carried in his pocket. He dared not sleep because of the cold. He counted in French and German and Italian to keep himself awake; he exercised as he lay in his narrow quarters. He told the unheeding ocean about the nature of the chemical bond.

When the stars came out, he sighted the end of his walking stick and tried to tell time by the constellations. He recited the periodic table of the elements. He grew more and more anxious, not for himself, since he knew he would eventually be found, but for Ava Helen, whom he could not tell that he was uncomfortable, but unharmed. He was chagrined by his predicament. When he was a graduate student in his early twenties, he had a rugged hour or so getting down from Strawberry Peak, in the Sierra Madre Mountains that overlook Pasadena. Since that time, he had hiked on a hundred mountain trails without so much as turning an ankle.

Dawn was a long time coming. By 9:00 A.M. on Sunday, the whole world heard the news, via radio, that Linus Pauling, lost all night in the rugged terrain of the Big Sur, California, had not yet been found. Sunday's midmorning news was tragic. Word came that Linus Pauling had been seen lying on a ledge at the bottom of a cliff. In about an hour, the searchers would reach him. The news was true, but most misleading, because someone changed the word *him* to *his body.* At about 9:45 A.M., one of the young volunteers sighted Pauling on his ledge. He went at once to tell the lieutenant sheriff, who passed on the news to the volunteer searchers, among whom was Barclay Kamb, Pauling's son-in-law. Ava Helen, waiting in the cabin, also heard the good news. In about an hour, she was told, her husband, who appeared to be unhurt, would be rescued from his narrow perch on the mountainside. Rescuers would take a stretcher and reach him from above—with ropes.

Linda Kamb, at home alone in Los Angeles with her infant twin sons, heard the false report of her father's death. So did Linus's sister, Pauline Ney, and Crellin and Lucy Pauling in San Francisco. Crellin Pauling and his wife were roused from sleep by friends who offered to baby-sit for them so they could go to the Ranch. They had not known of Pauling's being lost. There were tears of lamentation among Unitarians who had gathered for a Saturday night function at the church in Los Angeles.

In less than an hour, there was another "news flash." Rescuers had reached Pauling and warmed him with coats and hot coffee. The ledge where he had passed the night was only a mile from his cabin. With help, Pauling was able to clamber out of danger. He was smiling but very weary.

For a day or so, he was content to rest at home in Pasadena and read his mail, happy to be alive and well. Messages came pouring in from all over the world. He sorted the mail painstakingly and pasted the news releases into a scrapbook, along with letters and telegrams in order of arrival. As he replied to the messages, even to those that were no more than a calling card with "Thank God!" written on it, he carefully noted "Asw'd" in the right-hand corner of each missive in his neat script.

Letters came from strangers, acquaintances, old friends. A woman sent him a scapular to keep in his billfold to guard him from further dangers. An anonymous "Friend" sent him a Boy Scout whistle to serve the same purpose. "Don't tell," another correspondent wrote in block letters, "but I have visions, and I saw you on that ledge before you were found."

Letters came from every continent:

Both you and Mrs. Pauling have endeared yourself so very much to the Australian people.

I [an American college student] thank you so much for your wonderful textbooks, and many of us here hope that one day we will be fortunate enough to hear one of your lectures.

I wish to express my sympathy on the death of your husband. I met him only once and he had a warmth about him I will always remember.

We [Unitarians from his church] waited in the patio for the news and tears of joy and gratitude streamed down our checks when we found out that you were not dead, but alive.

One person after another said, "What a terrible day for the whole world!"

By noon we knew you were safe and my wife and I celebrated with an Old Fashioned.

I came home and found my wife in tears, word had gone around the hospital here that you were dead.

I told everyone at work that you would sit down and wait to be found, although the adolescent son of my employee got lost, then tried to walk out of the mountains in the night and fell off a cliff.

[From Tirupathi, India] Happy you are preserved for the benefit of mankind and the inspiration of younger scientists.

[From an English woman] That little house of yours looks isolated, for goodness sakes take care of yourself, get a couple of canines.

The television announcer who described your rescue said you looked happy, but spoke somewhat rapidly as if you were nervous. He doesn't know that your big problem in life is to move your mouth fast enough to keep up with your ideas.

Not everyone was happy to know that Linus Pauling had been found safe and sound. In April, the American Legion of Arcadia, California, started a campaign to save the world from Linus Pauling and Dr. Robert Hutchins, whom it named as "abettors of the Communist line."

Looking over the long list of charges printed in a local newspaper, Pauling noted that it had not done him justice. "Here they say that in November, 1957, I addressed a mass meeting and demonstration against nuclear weapons testing. It seems to me they are doing me an injustice. In the last four years I've spoken more than one hundred times a year against the tests, and against war."

Despite attacks from groups such as the American Legion, Pauling knew that the tide of opinion was turning in his direction, both in the United States and worldwide. By 1960, the Soviet Union, the United States, and Britain had all conceded the necessity of stopping atmospheric nuclear testing and attempts were being made to negotiate an agreement. Polls showed that the majority of Americans favored a test ban. Pauling had pitted his version of the truth against the established wisdom of the authorities and had been proved correct. He felt confident of his own abilities and ready to withstand any further criticism.

7

"A Consistent Pro-Soviet Bias," 1960–1966

Thanks to the activities of Pauling and others, public opinion in the United States was turning against the nuclear arms race with the Soviet Union. More and more prominent Americans were becoming involved in the peace movement. On May 19, 1960, the Greater New York Committee for a Sane Nuclear Policy organized the largest peace rally that had ever been held in the United States. Harry Belafonte and his chorus were volunteer entertainers, together with the Limelighters, Orson Bean, and Tom Poston. Speakers included Eleanor Roosevelt, Walter Reuther, Alfred Landon, Norman Cousins, Norman Thomas, Clarence Pickett, and Rabbi Israel Goldstein. Mike Nichols and Elaine May moved the mammoth crowd to half-tearful, half-smiling silence as they took them into a darkened bedroom where a pathetic, comic, misinformed husband and wife talked inanely about nuclear fallout. When Harry Belafonte kissed Eleanor Roosevelt on the cheek, the hundreds outside for whom there was no room inside Madison Square Garden heard the thunderous applause and speculated about what could have happened.

This antinuclear trend was looked upon with horror by Senator Thomas Dodd and others on the Subcommittee to Investigate the Administration of the Internal Security Act. On the eve of the Madison Square Garden rally, Senator Dodd announced that he was investigating Communist infiltration into the Committee for a Sane Nuclear Policy (SANE). He focused his attack on a volunteer who had worked on the preparations for the rally, claiming that the man was a Communist. Later it was discovered that the man had been denounced two years before by the *Daily Worker* as an anti-Communist. This information came too late for the board of New York SANE, however, who asked the man to resign in the interests of preventing bad publicity. The board also rescinded an invitation to Linus Pauling to speak at the rally in a last-minute effort to placate Senator Dodd, although Pauling's essay on nuclear fallout appeared in the program distributed at the rally.

Senator Dodd's actions had a devastating effect on SANE. Many prominent members and supporters withdrew from the organization to protest its caving in to a witch hunt. The national organization cut itself off from the New York City chapter. The New York chapter advised thirty-seven members of the organization who had received subpoenas to appear before Dodd's committee to choose a closed rather than an open hearing in order to minimize publicity. Dodd was greatly encouraged and eager to get on with the hearings, which could discredit the peace movement and force it to undergo an anti-Communist purge if it wanted to survive at all.

Pauling was dismayed at SANE's action but continued his own activities. On June 18, 1960, he spoke in Washington, D.C., at a meeting of the Women's International League for Peace and Freedom. After his speech, he was surrounded by friendly members of the audience who wanted to talk to him. One woman gave him a poem; another gave him a musical score. He accepted newspaper clippings and mimeographed material "to read later." That evening, while he was going through his papers to find a schedule indicating when he was to appear on a radio program, he found a subpoena, signed by Senator James Eastland, commanding him to appear in Washington, D.C., on June 20, to give witness as to Communist infiltration of the campaign against nuclear testing.

Pauling immediately went on the offensive. His old friend, Abraham Lincoln Wirin, counsel for the Southern California Chapter of the American Civil Liberties Union, was recuperating from surgery, but he immediately traveled to Washington to act as Pauling's counsel. The subpoena did not specify whether the hearing would be open or closed, but when they arrived they discovered that a closed hearing was planned. Pauling immediately sent a telegram to the counsel for the committee requesting that the hearing be open so that all the testimony, and not just selected portions of it, would be available to the public.

An open hearing was granted, and the proceedings began June 21, 1960. Senator Dodd and the subcommittee's attorney, J. G. Sourwine, had had ample time to read the information sent to them about the petition campaign before the hearing. They had had a complete photostatic copy of the entire petition, and of a press release, two and a half months before the hearing. The copy of the petition that contained the list of the entire 11,021 scientists lay on Dodd's desk near at hand during the hearing.

Nobel Prize winners among the signers were listed first, in alphabetical order, and according to their field of eminence. Alphabetized lists of members of the U.S. National Academy of Sciences and Fellows of the Royal Society of London followed. The remaining names were listed alphabetically, with subdivisions that designated their country, and, in the larger countries, the city where they were employed. A visitor to the UN, a delegate, or a reporter—or anyone who was interested—could quickly discover who had signed the petition from a given country or city. He could discover that Peter Brian Medawar, Department of Zoology, University College, London, who later received the Nobel Prize, had signed Pauling's petition, as had Hideki Yukawa, a physicist at the University of Tokyo, who also became a Nobel Laureate.

Readers interested in numbers could quickly learn that the first report to the UN, submitted in January 1958, contained the signature of Linus Pauling and 9,234 other scientists. The second report, listing 1,803 names in similar fashion, was mailed to the UN in July 1958. Although the subcommittee based its attack on Pauling on the premise that some of these names were not "actual names," they had not bothered to check the list's authenticity. As the hearings progressed, Pauling found himself lecturing senators on how they should have gone about investigating him, on how easy it would have been for them to write to some of the scientists or to all of them. He had done this study with very little help; they had clerical help and government funds at their command.

If the senators had been serious about wanting to gather information about the petition campaign, the expense of holding the hearings could have been spared the American taxpayer. The information they sought was readily available. The eleven thousand names had been on file for two and a half years at the UN; press releases had been sent all over the world; Pauling had described the gathering of signatures in his book *No More War!,* a copy of which he had given to every American senator in 1958; and hundreds of magazine and newspaper articles had been written about the petition. In all that time, not one scientist had protested that his name had been used without his permission. No one had questioned the authenticity of the signatures until Senator Dodd did so.

The real purpose of the hearings was to ridicule participants in the peace movement publicly, stigmatize them as subversives, and thus discourage others from supporting their cause. By demanding an open hearing and using it as a vehicle to expound his beliefs, Pauling cut to the heart of the matter. Under such circumstances, many witnesses refused to talk at all or responded with guarded replies after whispered consultation with their counsel. Pauling responded to questions with a lecture. He answered rapidly without premeditation. He often set the spectators laughing, and Senator Dodd was frequently driven to threatening to clear the hearing room. "This is not a circus!" Dodd shouted angrily.

When Pauling discovered that Senator Dodd and Attorney Sourwine were bent on presenting him to the world public as a cheat and a fraud, he turned very quickly into the stern, watchful professor. He always worked hard, but he outdid himself during the hearings. He worked to improve the senators' ethics as diligently as he worked to enlighten their minds. They held honorable positions, and he felt that they disgraced their positions. It was his duty as a citizen to make them see the error of their ways. He noted their every delinquency, exposed it, analyzed it, and saw that his statements were accurately recorded in the *Congressional Record of Hearings.* He checked not only the galley proofs, but also the stenographic records to be sure that the hearings were reported accurately. He corrected spelling and punctuation. The title chosen for the printed account of hearings by Senator Dodd and his committee was *Communist Infiltration and Use of Pressure Groups, Hearing before the Subcommittee to Investigate the Administration of the Internal Security Act and other Internal Security Laws of the Committee of the Judiciary, United States Senate, Eighty-sixth Congress, Second Session, Testimony of Dr. Linus Pauling, June 21, 1960.*

Pauling discovered this title in the galley proofs and insisted that he had no information whatsoever to give on this designated topic. The title was consequently changed to *Testimony of Dr. Linus Pauling.*

The debate between Pauling and Dodd began in the press before the hearing started, and the public anticipated an even more exciting encounter than that between Teller and Pauling. Senator Dodd had told reporters, "Pauling has displayed a consistent pro-Soviet bias."

Pauling had replied: "If Senator Dodd charges me with Communist activities away from the sanctuary of Congress, I will sue him for libel."

During the first open session, both the senator and the professor were on their best behavior. Pauling had previously consented to appear at a brief closed session, at which arrangements to hear him in open session were made. Senator Dodd agreed that subsequent sessions would be open to the press and the public. Dodd, during the first open session, made a long discursive statement as to the reasons for the hearing. Pauling was given an opportunity to reply to the statement.

During the afternoon session, Senator Dodd said that he had read only a part of his statement during the morning but that he would not finish it. "The record will contain the entire statement if you [Mr. Wirin] do not have an objection. I wanted to have an opportunity to read it, but I do not think that you will find anything objectionable." Attorney Wirin consented, but *only if* he and Pauling would have an opportunity to read the omitted words before they were printed. He was not told that Dodd's entire statement, the sixteen hundred omitted words and all, had already been sent to the press, even before the partial version was read.

The omitted passages that Senator Dodd chose not to read in Pauling's presence would have elicited strong protest from both Pauling and Wirin, if they had read or heard them. The senator had avoided saying in Pauling's presence:

> Furthermore, our interest in Pauling's petition is justified by Dr. Pauling's long record of service to Communist causes and objectives, many of them related in no way to his special field of science.

Senator Dodd also chose not to question the validity of the petition in Pauling's presence. In the statement sent to the press and the printer he did so. He had written that he wanted to be certain that the names Pauling had filed with the United Nations were "actual names." The press and public had the right to assume that Pauling had heard these accusations and had made no protest. If the press passed this assumption on to the public, Pauling would be thought guilty since he did not try to defend himself.

Since neither Pauling nor Wirin had as yet learned of these serious accusations, they were unprepared for what was to happen during the first open session. There was nothing in Senator Dodd's welcoming address to indicate that he hoped to brand Pauling as a liar and a fraud before the day was over. He said:

> Now, Dr. Pauling, let me say we are pleased to have you here this morning. We regard you as an eminent scientist, a man who has made a great contribution for the betterment of mankind. . . . I hope that you will feel, as we want you to feel, and as you should feel, that there is nothing hostile about our presence here, nothing hostile to you at all.

A few minutes later, Senator Dodd entered as an exhibit what became known to the press as the "Macomber Letter." The senator asked each of the other senators present (Keating, Cotton, and Johnston) to read it. They did so, frowning ominously. When Pauling and Wirin were given the letter to read, they looked puzzled and intent. Wirin scowled. Pauling stopped smiling. Tension in the hearing room, especially among Pauling's friends and well-wishers, became almost unbearable. The press was alert, intent.

The letter from William B. Macomber, Jr., assistant secretary of state, questioned the validity of the Pauling petition. Macomber had reported, "Only a partial list was communicated to the Secretary General of the United Nations." This was a more serious accusation. Pauling's integrity was being questioned.

Pauling was placed in the uncomfortable position of having to blame the United Nations, an organization he admired and wanted to see assume a greater role in world government. He had personally, in the presence of his wife, placed the first petition listing 9,235 signers in the hands of Mr. Dag Hammarskjold on January 15, 1958. The second list of almost 2,000 names, bringing the total to 11,021, had been delivered by them to the United Nations a few months later. He could not understand why Mr. Macomber had not found the complete petition—unless the UN had lost the second list of petitioners.

If the petition were lost, or mislaid, he had a carbon copy of the document at home, as well as a photographic facsimile of the petition, and the original signatures, all in a safe in Pasadena. The press reports, he added amiably, did not give all the names since no newspaper would print a list of eleven thousand names. He and his wife had selected names for the press report using their own best judgment. The tension in the courtroom subsided. Pauling was smiling and relaxed. So were the spectators.

Dodd admitted that Macomber might be in error. Did Pauling know why? Pauling did not think it was his responsibility to know why. Perhaps Senator Dodd could find out for himself.

Attorney Sourwine, speaking out of context a half-hour later, exclaimed, "You will note that in this press release, there is no list of Soviet signers. Were they eliminated for any particular reason?"

This was a clever move by the attorney. He raised doubt once more about Pauling's integrity, and also brought in the name of the Soviet Union and connected it with Pauling in some unspecified way that was never defined during twenty minutes of rigorous interrogation about the matter. Pauling had no recourse but to blame the United Nations. The press release should contain the names of Soviet signers. If they were not listed immediately after the names of the Fellows of the Royal Society, someone at the UN must have abridged the list for some reason.

The farcical presentation of the "Macomber Letter" and the missing Soviet list came to an abrupt halt when Pauling discovered that the Soviet list was just where he said it was, and that Senator Dodd had the complete petition on his desk, a photostatic copy sent to him by the UN listing all 11,021 names, when he introduced the "Macomber Letter" in an attempt to suggest that such a list did not exist. Pauling mentioned his discovery at the hearing, but there was no response from the senator or his attorney, who went on to other matters, ignoring the discovery of their trickery.

Pauling and Wirin were alerted. Senator Dodd, evidently, was not to be trusted. Thereafter, they critically examined all statements made. They asked for the galley proof to check the content of the sixteen hundred words Senator Dodd had omitted in his opening statement and protested their inclusion in the printed record. The protests were unheeded. Pauling was finally permitted to introduce a sixteen-hundred-word protest at an October hearing. Senator Dodd had put sixteen hundred derogatory words into the record that Pauling had not heard, although the record falsely indicated they had been spoken in Pauling's presence. Pauling was permitted to enter a protest of not more than sixteen hundred words—but not until October 11.

Dodd continued to express skepticism about the genuineness of the signatures and asked Pauling to produce the "real" signatures in Washington for the subcommittee's inspection. This Pauling was most eager to do, since the question of their number and existence had been raised. He consented to submit a list of all persons, at home and abroad, to whom he had sent the petition. He said that no one could be blamed for receiving a letter. The senators also wanted to know who had helped him in any way, worked the mimeograph machine, helped to mail letters, gathered petitions and returned them. If he had such a list, they could examine it to see whether or not in their opinion these persons were also part of some subversive organization.

This Pauling adamantly refused to do. He believed in the right to petition:

> The circulating of petitions is an important part of our democratic process. If it were to be abolished or greatly inhibited, our Nation would have made a step toward deterioration—perhaps toward a state of dictatorship, a police state.

He also refused to give the names of those who had helped to gather petitions:

> As a matter of conscience, I should not subject people to reprisals for circulating petitions. This is a part of the democratic process.

Senator Dodd was shocked at the idea that any loyal citizen would mind being subpoenaed to appear before the subcommittee. What was this talk of reprisal? Although Pauling was threatened, obliquely, with imprisonment, he continued to refuse:

> I am responsible for my actions, and I wrote the petition and I sent it out to people, asking that they get signatures to it. I even selected the many people to whom these petitions were to be sent. I think that my reputation and example may well have led many younger people to work for peace in this way. My conscience does not allow me to protect myself by sacrificing these idealistic and hopeful people, and I am not going to do it. As a matter of conscience, as a matter of principle, as a matter of morality, I have decided that I shall not conform to the request of this subcommittee.

He was eager to tell them more about his own activities as a gatherer of signatures. He reminded them that Fulton Lewis, Jr., had estimated that gathering the petitions must have cost him $100,000, or approximately $10 a signature. His estimate was $.03 a signature. He and his wife had spent about $250 sending the mail that went to foreign scientists. The spectators applauded and laughed when Pauling said that he and Ava Helen had been happy to make this "small financial contribution to the solution of world problems." Dodd threatened once more to clear the hearing room.

Senator Dodd had the privilege of making the final remarks that closed the stormy second session, on the afternoon of June 21, 1960. Pauling was "ordered and directed" to appear once more before the subcommittee. He was to bring the "signatures or purported signatures," a list of those to whom petitions were sent, and the information the subcommittee needed in order to know who helped him gather the signatures. Pauling left Washington, D.C., faced with the prospect of a citation for contempt of court, and possible imprisonment.

As usual, the world press supported Pauling. Citizens in other countries were fully informed about the "Macomber Letter" and of Pauling's refusal to name those who had helped him. Since the need for a nuclear test ban was not a debatable issue to most Europeans, the comments were often more direct and explicit than in most U.S. newspapers, which were more eager to avoid taking sides until the issue had been settled.

When the hearings resumed on October 11, 1960, Pauling presented the original signatures. The twelve thousand pages full of signatures were bound into three great volumes. Senator Dodd was now very receptive, and pathetically eager to get the impressive presentation over with as little ado made about it as possible, but Pauling was proud of these three volumes and wanted to talk about them. The spectators in the crowded hearing room were as eager to hear and to see.

There were additional signatures in the scrapbooks of scientists who had sent them too late to be included in the petition. The petition from a lone scientist in Laos had fallen behind a heater and had not been included; the total number of countries represented therefore was not forty-nine, as he had said, but fifty. He was fairly sure that a few original signatures were missing. Dodd repeatedly assured him the subcommittee was quite satisfied with the evidence as presented.

The copies of the translations of petitions into different languages, including Thai and Russian, had no significance, Pauling leisurely told the unresponsive senators. He had included them only because it gave him pleasure to do so. He turned the pages, enjoying his handiwork. While he prolonged the presentation, the senators chafed under the strain of being confronted with evidence of its false accusations.

Reporters and spectators in the hearing room were eager to hear what Pauling would say when he was asked for other documents that would reveal the names of those who had helped him. If he did not have them, would he be cited for contempt of Congress? And imprisoned?

When Senator Dodd finally made the delayed request for these names, Pauling began to bargain. He would not answer the senator until he was assured that he could make a statement as long as Senator Dodd's preliminary remarks had been in June. Senator Dodd indignantly refused and said he would "not be blackmailed." In the heat of the ensuing argument, Pauling made it quite clear that his answer to Senator Dodd's request for the names of his helpers was an unequivocal no.

"Very well," the senator said, and told him to say whatever he wanted to say. For nearly an hour, the senators listened while Pauling itemized the shortcomings of the senators and lectured them earnestly about the responsibility of being senators, about their moral bankruptcy.

He said that he felt that he was not fighting for himself alone when he objected to injustices done him. "I must say that in making the fight against what I consider to be injustice done to me by this subcommittee, I am fighting also for other people who may not be as able as I to make a fight for themselves." He said he wanted very much to have laws passed to make this kind of investigation illegal in the future.

He could not conceal his anger over being lied to and lied about. He said in reference to the presentation of the "Macomber Letter":

> Moreover, it was known to you (Senator Dodd) at the time this letter was introduced into the hearing and during the twenty minutes when I was being interrogated about the "incomplete" list of names of signers of the bomb-test petition that there rested on your desk, in front of me but not visible to me, a photostatic copy of the complete list of signers that I had turned over to the Secretary General of the United Nations. I know, because the complete list was at this later time brought out and handed to me for identification. . . . I am greatly disturbed to be forced in this way to believe that a committee of the Senate of the United States would be guilty of an immoral action of this sort. But the facts do force me to believe it.

Pauling introduced into the records of the second hearing his protest against the statement Senator Dodd had made in June about his "long record of service to Communist causes and objectives":

> I have not served Communist causes and objectives, and I am indignant that the subcommittee should accuse me of doing so. All of my actions have been characterized by independence. All of my life I have striven to discover the truth and to behave in a way compatible to the greatest possible extent with the highest moral and ethical principles. In my actions about social and political questions and the great question of preventing war and preserving the peace in the world I have served no one except the whole of humanity, but have made up my own mind about every issue and have always taken those actions that in my opinion were the right ones. I have striven to do my duty as a loyal American citizen, in

> ways indicated by and corresponding to the Constitution of the United States.

The chief counsel, Pauling said, had questioned him for twenty minutes about the "missing list of Soviet signers."

> I was not given an opportunity to examine the press release at that time. I accepted this statement and said: "I am sure that a mistake had been made."
>
> The photostatic copy of the press release was introduced in evidence then and printed in the printed record of the hearing, and the list of the 216 Soviet signers was in it, and was in it at the time. . . . This document . . . had been in the hands of the subcommittee for two months and three weeks. . . . I find it difficult to believe that during this period . . . the subcommittee was not able to discover . . . this list of 216 names.

He tried to impress upon them the responsibility of being senators:

> I must say that I have a great respect for the Senate of the United States. I have been brought up in such a way, my boyhood in the western United States and my whole life has been of such a nature, as to lead me to believe that we live in a great country, the greatest country in the world, and I believe in our democratic form of government.
>
> I have looked forward to the experience of appearing under subpoena before a congressional committee for the first time (he had once before appeared before a committee as a voluntary witness) which was the 21 of June this year. I had thought that this would be an opportunity for me to see the way in which an investigatory committee . . . discovers the truth about important matters. . . . I cannot conceal that it has been a great disappointment to me.

Senator Dodd made no response to Pauling's exhortation. Pauling's attorney A. L. Wirin and Senator Dodd conferred on legal matters pertaining to the petition for a writ of certiorari Wirin had filed in Pauling's behalf between hearings, asking that Pauling not be compelled to attend a second hearing. The Supreme Court had not yet made its statement when the appointed time for the second hearing had arrived, and Pauling had traveled to Washington, D.C., rather than complicate legal issues further.

Attorney Sourwine then gave as evidence of Pauling's Communist activities in relation to the nuclear test ban an advertisement by seven foreign language publications, which he said, "essentially constitute the Communist foreign language press of this country." They, too, favored the test ban. This proved, he implied, that Pauling believed as Communists believed, and therefore was a Communist.

It was increasingly evident that Pauling was not going to be cited for contempt of Congress. The hearings went on and on, but no further mention was made of the information Pauling had refused to give about persons who had helped him. The subcommittee concentrated instead on trying to link Pauling with specific persons and specific organizations they designated as Communist. Much of the material, like that concerning the foreign language press, had no realistic relationship to Pauling whatsoever, and the crowd dwindled to a devoted—or conscientious few.

Sourwine asked Pauling whether or not he had ever signed a petition in behalf of anyone who was a Communist. Pauling replied that none of the persons in behalf of whom he signed was, as far as he knew, a Communist. "I do remember signing a statement urging that Gerald H. K. Smith be allowed to speak in a public school in Los Angeles, and I feel sure, reasonably sure, that he is not a Communist." Smith was an American fascist leader; Pauling had supported his right to speak purely on civil liberties grounds.

Pauling was an exasperating subject to interview. There were times when the subcommittee must have been certain that it had made a strong point against Pauling, but he seemed to take pride in activities they deemed shameful. He was eager to talk about international peace conferences he had attended. He was not at all afraid to say that he knew various foreign scientists who were said to be Communists or to be sympathetic to Communism. He was proud to be a member of the Academy of Sciences of the Soviet Union. He had not read much of Marx, was not a Marxist, but he would like to read Marx some day—when he had time. He would not be influenced to say that scientists in any nation were completely free. The members of the French Academy of Science, he said, might also be called up by some investigating committee and quizzed. He doubted, however, that the Russian scientists were as free as the French scientists from political domination.

> You know, I think that there is a lot that can be done for civil liberties and civil rights in every country, including our own. But I think there is probably more that can be done in the USSR than in France or the United States. If I ever get the opportunity to do something for civil liberties and civil rights in the USSR, I shall do it.

By midafternoon, the crowd had dwindled. The students who had packed the room were gone. The spectators who were left were pro-Pauling. They laughed and applauded when, in denying that he had followed the Communist line, Dr. Pauling said that some people accused the Communists of following the Pauling line. A. L. Wirin was stifling his yawns. Ava Helen Pauling, readily identifiable in her gay blue hat with the red and white flowers, remained attentive and kept on taking notes for her husband.

As the hearing grew tedious, more and more men and women drifted away from the hearing room. Some preferred to stand and talk together in the hall outside. They hurried back in when word was passed around that Sourwine seemed to have some last card up his sleeve, some final document linking Pauling to a Communist group at Washington University, St. Louis, where the first petition had been initiated.

Sourwine's surprise turned out to be Pauling's acquaintance with the eminent physical chemist Martin Kamen, best known as the discoverer of carbon 14, who had been one of the twenty-seven original signers of the petition. Pauling readily confirmed that he knew Kamen, who was then employed at Washington University. Sourwine read a lengthy report about Kamen's alleged espionage activities in connection with the atom bomb, which had been published by the House Un-American Activities Committee in 1948.

Kamen has since told the whole story of his false accusation by the committee in a moving book, *Radiant Science, Dark Politics.* He had been taken to this country from Russia at the age of three months. As an outstanding scientist, he had received an important job in a war industry. On July 1, 1944, he was seen at Bernstein's Fish Grotto in San

Francisco having dinner with a Soviet vice-consul. Government agents had apparently testified that they believed he gave classified information to the Soviet consul that night. There was no basis for the accusation; the root of it was that because of his knowledge of physics Kamen had inadvertently implied certain information about the American nuclear program that was not yet public knowledge.

Sourwine asked whether Pauling knew of this incident. Pauling did.

> I remember, too, that he was completely cleared of these charges and won a libel suit against a Chicago newspaper. He was completely cleared of these charges, and to introduce the statement that you have introduced without telling the rest of the history seems to me to be an effort at a smear of (him) and perhaps of me by association.

Kamen had also been cleared by the court that gave him back the passport that had been denied him. Pauling had contributed a small sum, less than he would have liked, he said, to the fund raised to help Kamen regain his passport. "He is a respected citizen, one of the victims of the sort of oppression that goes on."

What had happened to this bright, young chemist was evidence, Pauling said, of what might have happened to other young scientists if Pauling had consented to name his assistants in gathering signatures for his petition. Sourwine dropped the subject abruptly and soon made another absurdly farfetched attempt to prove Pauling guilty by association.

He asked whether Pauling had had any dealings with Carl Marzani and Alexander Munsell, both of whom "have records as members of the Communist party."

Pauling had heard of neither of them.

Sourwine explained (while Pauling's smile grew wider and wider) that Pauling's book, *No More War!,* published by Dodd, Mead ("A good firm," said Senator Dodd), a respectable old firm whose fortune was founded on the Elsie Dinsmore series, had sold the rights to publish a book-club edition of Pauling's book to Marzani and Munsell.

Pauling explained that he had no part in the business arrangements involved, and that he had never heard of these two gentlemen until Sourwine had mentioned them.

The hearing dragged on in a desultory fashion until the subject of the peace demonstration in San Francisco on May 13 and 14 was introduced. Pauling, who had addressed the peace marchers, gave his version of the affair in concise detail. He said that he had been asked to speak by a member of the American Friends Service Committee. He was asked whether he knew who had organized the peace march. The implication was that it had been Communists. Pauling said it was organized by the American Friends Service Committee, the Fellowship of Reconciliation, the American Federation of Labor (AFL) Labor Council, and the San Francisco Peace Congress. He gave specific detailed information about happenings during those two days.

Did he know that J. Edgar Hoover had characterized the incidents at San Francisco as Communist inspired? Pauling said that he was not interested in what Mr. Hoover had to say about these demonstrations.

Senator Dodd, who said very little during the sessions of the second hearing and at times protested evidence Sourwine introduced as irrelevant, was irked by this remark of Pauling's about J. Edgar Hoover and ordered that a twelve-page statement by Hoover of the San Francisco "riot" be read into the permanent record. The subcommittee was interested in what Hoover had to say, Dodd grumbled, even if Pauling was not.

A resume of world peace activities from 1949 to 1959 was also read into the record. The subcommittee and Pauling had difficulty in understanding each other's attitude toward this rather competent survey. Pauling looked upon it with pride and so did they but for very different reasons. Although he thought the subcommittee revealed its bias in the wording at times, he made no objection to its being included in the records. He wanted very much to expand upon some of the items. Not enough emphasis, he felt, had been put on the Pugwash Conferences. The inference that they were Communist dominated was false:

> The Pugwash meetings of scientists are not at all dominated by either the East or the West; perhaps a bit by the West, because the number of scientists from Soviet countries is considerably smaller than from Western countries. . . . I am glad the Pugwash conferences are going on. They are an important part of the effort being made now to find some solution to

the terrible problems the world faces. And I do not mind having this introduced into the record.

The professor was willing to lecture at length on the world peace movement, but the disgruntled senators were most unwilling to listen. The hearings were adjourned.

In 1961, the Dodd Committee issued a fifty-nine-page summary report, *Testimony of Linus Pauling.* Among the subsections are "Pauling's Role in the International Peace Offensive," "Pauling's Record in Regard to Associations with Individuals with Known Communist Records," and "Pauling's Opposition to Anti-Communist Legislation and Measures."

The Dodd Committee had the last damaging word. There was nothing Pauling could do to prevent this great mass of misleading material from being made a part of the permanent *Congressional Record.* The summary report included possibly libelous accusations that had not been made at the hearing. It also repeated accusations to which Pauling had repeatedly replied and did not quote his replies. Once more he was presented as a strong force in a Communist conspiracy. One of the new accusations was that Pauling was disloyal because he had signed a petition urging that the president of the United States meet with the premier of the Soviet Union in a face-to-face conference on the use of nuclear power.

In October 1961, Pauling sent a long letter to both Premier Nikita Khrushchev and President John F. Kennedy asking that they influence their countries to refrain from making further bomb tests. The reply Pauling received from his own government was the perfunctory statement of policy sent through regular channels to other citizens who made similar statements.

Khrushchev's reply, however, was personal, couched in his own idiom, and signed by him. As such, it is of considerable importance as a historical document. The Russian premier was constantly defensive in his reply: "Try to understand, dear Mr. Pauling, what the Soviet Union would be like if it continued to refrain, as if nothing at all has happened, from taking additional measures to perfect nuclear weapons while NATO powers are responding with threats to its proposal that a German peace treaty be concluded."[1]

In late April 1962, Pauling was again in the news because he picketed the White House shortly before he was a guest there. President John F. Kennedy and his wife invited Nobel Prize winners in the United States and one hundred four other American and foreign scholars to the White House for dinner on April 29, 1962. The public knew the names of the literary figures who were invited: James T. Farrell, Robert Frost, John dos Passos, Pearl Buck, Katherine Anne Porter, Van Wyck Brooks. William Faulkner declined and gave no reason.

Very few scientists were invited, and two of those who accepted were controversial figures. Senator Karl E. Mundt said that Dr. J. Robert Oppenheimer should not be included. Even if Oppenheimer were invited, he should have known better than to accept, the senator declared. Pauling's invitation was also criticized.

There were those of his admirers who thought that Pauling should have declined the invitation as did Faulkner, but Pauling did not agree with them. He believed in friendly communication between those in disagreement and was pleased with the unprecedented recognition of scholarship and talent by a president and his wife.

Several day-long "Ban the Bomb" peace demonstrations outside the White House had been scheduled for the period that included the dinner. Pauling saw no incongruity in picketing the White House before dining with its occupants. Clarence Pickett, the venerable and genial Quaker who represented the American Friends Service Committee when it, together with its English counterpart, was given the Nobel Peace Prize, felt as Pauling did. They both accepted both invitations: to picket and to dine.

Jacqueline Kennedy was not perturbed by her picketing guests. She smilingly told Pauling that her daughter, Caroline, had noticed the picket line and had asked, "What's daddy done wrong now?"

In April 1961, Pauling went to Honolulu. For the first time since 1951, he was asked to speak at the University of Hawaii. On the evening of April 7, he addressed the university students on the subject "Scientists and Public Affairs." The Political Affairs Club at the university was his sponsor. The student paper, *Ka Leo,* told the saga of his 1951 repudiation by the university in scholarly detail with quotations from what was, to the students, a very long time ago. Pauling was the hero of the saga, the Ulysses returned. The students met to hear Pauling in Bilger Hall, named for Dr. Leonora Bilger, who had suggested in 1951 that Pauling be asked to lecture at the dedication exercises that were postponed indefinitely because of the unanimous disapproval of the nine regents. Dr. Bilger was Pauling's staunch defender in the 1951 period described by *Ka Leo* as one of "much concern and anxiety on campus."

Many Islanders took an almost personal pride in Pauling's Nobel Prize. To them he was a quasi Islander because he had been a frequent visitor and had grandchildren on the island. Linus Pauling, Jr., the psychiatrist, was now a respected member of the island's professional community.

Pauling used the occasion of his visit to Honolulu to announce a new theory that he had developed, the hydrate microcrystal theory of general anesthesia. This was unusual at this stage of his life, for most of his public speaking had been on nuclear fallout or other topics of current public interest rather than on his own contributions. A new theory had, however, come to him one morning while he was casually reading an article in a scientific journal.

Nothing was further from his mind that day than creating a theory of the molecular mechanism of general anesthesia. The manner in which he arrived at the theory is illustrative of what Pauling called his "stochastic method," a kind of informed guessing wherein the researcher, because of his openness of mind, curiosity, and tremendous stack of factual knowledge, finds himself having an idea that no one has ever had before. This discovery, described in detail in the preceding chapter, occurred when Pauling read a paper on the crystalline hydrate structure of an alkylammonium salt, which had been found to stabilize cagelike structures of water molecules, converting liquid water into an icelike framework consisting of twelve-sided chambers.

To this information, Pauling added his own knowledge that the melting point of these structures could be raised from 25° C to 37° C if the chambers were not empty but were occupied by molecules small enough to fit into the cages, yet large enough almost to fill them.

At this point, he made the crucial comparison—these alkylammonium salts resembled the amino acid chain of proteins normally found in the brain. At this instant the jigsaw puzzle he was creating began to fall rapidly into place.

Assuming that the structuring of the water molecules around the proteins of the brain might interfere with the motion of ions or electrically charged protein side chains that normally contribute to the electrical oscillations of the brain involved in consciousness, it could be expected that the amplitude of these oscillations might be reduced enough to reduce consciousness.

This new theory could be used to explain several phenomena such as why deep sea divers become unconscious while breathing air under high pressure, and why the rare gas xenon is in effect anesthetic. Unfortunately, attempts by several of Pauling's students and associates to confirm the theory experimentally were unsuccessful. The stochastic method is better for raising ideas and stimulating research than for resolving issues, and the question of the mechanisms of anesthesia remained unanswered.

In Pauling's view, this incident illustrated the way he frequently used his unconscious mind in making discoveries:

> Some years ago I decided that I had been making use of my unconscious in a well-defined way. In attacking a difficult new problem I might work for several days at my desk, making appropriate calculations and trying to find a solution to the problem. I developed the habit of thinking about a problem as I lay in bed, waiting to go to sleep. I might think about the same problem for several nights in succession, while I was reading or making calculations about it during the day. Then I would stop working on the problem, and, after a while, stop thinking about it in the period before going to sleep. Some weeks or months, or even years might go by, and then, suddenly, an idea that represented a solution to the problem would burst into my consciousness.[2]

Pauling titled his speech on this discovery "The Genesis of the Concept of Molecular Disease" and frequently cited the discovery of the idea of molecular disease as one of his accomplishments. While this idea no doubt occurred to him independently, his assumption of priority overlooks the important work of Archibald Garrod, who, in the early part of this century, demonstrated that the human disease alkaptonuria was caused by a genetic inability to metabolize a form of acetic acid.[3]

Over the years, Pauling had tried to work closely with his son Peter, who was trained as a physicist. This had led to some interpersonal conflict. In 1963, Peter proposed that Pauling write a high school chemistry textbook "in collaboration with some flunkey to do the hackwork."[4] Peter offered to pay his father $10,000 for the manuscript, complete with illustrations, which he would then publish with his own one-man publishing company.

Pauling was not impressed with this offer. He thought that a high school text was worth much more; it might make as much as a million dollars. He thought the time was ripe for a high school text, which would complement successful texts that he had written for the college market. However, he did not want to write it. In a lengthy letter, he explained his reasons:

> The fact is that I don't want to be bothered to write a high school text, and I doubt that I shall do so. First, it is very hard to find a satisfactory collaborator. I have tried two or three, without success. My experience is that it is about as easy to do something myself as to get somebody else to do it, and if I do it myself I am satisfied with it, whereas if I get somebody else to do it I am not."[5]

When Peter pursued the issue, Pauling explained:

> During the last couple of years I have thought a great deal about the matter of alternative ways in which I can spend the remaining years of my life. There are many things that I want to do—so many that I have to decide among them. I want to make the decisions, in cooperation with Mama, in such a way as to be most satisfying to her and to me. I do not know yet what the decisions will be. I am glad that my circumstances are such that I do not need to take on the uninteresting chore of writing a high school text. I am going to try to have the decisions made in such a way as to give Mama the most pleasure and satisfaction during the remaining years of her life, and also to give me as much satisfaction as possible. . . . In answer to your question, I think that Mama and I need to be repaid, from now on, for any investment of time and energy that we make—we need to be repaid in happiness, not in money. Mama and I both send you our love.[6]

In response to Peter's offer, Pauling suggested that Peter write the book himself with his father as nominal coauthor. He offered to advance Peter $10,000 a year for two years while he wrote the manuscript and to let him have the majority of the royalties if he did the majority of the work. This was a potentially lucrative offer, since a high school chemistry book by Linus and Peter Pauling might have done very well.

Although Peter's doctorate was in physics, not chemistry, he undoubtedly knew enough chemistry to write a high school text. Nevertheless, he rejected the offer bitterly:

> I know it would be a good thing for me to take whatever time is necessary out in order to write a book with your help which would repay me so well. But I cannot, I am not a chemist, do not know how to write books and cannot be a second you.[7]

Pauling had, in effect, asked Peter to take on the role Peter had described as "some flunkey to do the hackwork." Emotionally, he was unable to play this role, however much sense it might make financially.

Peter was having personal difficulties of his own, quite apart from his relationship with his parents. His first marriage ended in divorce, and he had problems with depression. His angry outbursts reflected the low points in his feelings; when he was feeling better his feelings toward his parents, and especially his mother, were quite affectionate. The correspondence between them that is preserved in the archives is generally affectionate and supportive. Even at the worst of times, Linus and Ava Helen's correspondence was always supportive.

After his divorce, Ava Helen counseled him on relationships, suggesting that he would be happier if he could commit himself to a meaningful relationship with another woman. Peter acknowledged that he was "almost certain to marry again, but when I do not know."[8] Peter went through a period of heavy drinking followed by what he called a "post-booze manic phase" and a "post-manic post-booze depressed phase."[9] Throughout these difficulties, Linus and Ava Helen did what they could to be supportive, and Peter wrote later to express his appreciation for their support and understanding. There was also a lot of correspondence at happier times, with talk about buying cars, arranging visits, and other ordinary family business.

There is also a great deal of correspondence from Linus with practical advice about scientific matters, getting published, getting funding, and so on. Peter sometimes invited this response with letters asking for help and advice. He also asked for and received financial help when he needed it.

Peter worked through the difficult times in his life, with support from his parents and others, and remarried successfully. He continued living in London, visiting and corresponding with his parents, but naturally not seeing them on a daily basis. He pursued his career largely independently of his father. They never did the high school text, but Peter did appear as coauthor on the 1975 revision of Linus's *College Chemistry.*

Linus and Ava Helen Pauling were at the Ranch on the day when two events occurred that were of tremendous importance to them both. On October 10, 1963, the same day the partial nuclear test ban treaty was signed by the three nuclear powers, the Norwegian Peace Price Committee announced that Linus Pauling had been awarded the Nobel Peace Prize for 1962. No suitable candidate had been found by the committee during the previous year. Since the selection process used by the Nobel Committee of the Norwegian Parliament is complex and requires much study and the cooperation of many individuals, it seemed likely that the committee had been waiting for the partial test ban treaty to be signed with the intention of eventually giving the 1962 award to Pauling. He had, however, no inkling that such plans were afoot.

The telegram arrived at the house on Pierpont Street in Pasadena, and no one knew quite how to reach the Paulings. There was no telephone at the Ranch. The ranger, who had the only phone in the area, obligingly gave the news to his neighbors.

While Linda and her husband were waiting for their parents to get home from Big Sur, they occupied themselves and the twins by decorating the living room with signs. LP + AV + PAX = 2 NOBELS. PEACENIK PAULING PICKED FOR PAX PRIX. The Paulings arrived; the photographers were waiting. Pauling obligingly dropped on his knees on the Oriental rug in the living room and his excited twin grandsons clambered on his back to play "horsey." The telephone rang constantly, and the telegrams lay in heaps on the table.

Once more all the Pauling children accompanied their father and mother to see him receive his prize—at Oslo—and to hear him give his acceptance speech. Crellin and his wife traveled from the University of Washington, where he was doing graduate work in genetics. Peter and his wife traveled from the University of London, and the Barclay Kambs from Altadena, California. This time, the oldest of the thirteen Pauling grandsons, Linus Pauling III, the black-eyed, broad-shouldered, husky twelve-year-old son of Linus Pauling, Jr., was old enough to attend. His mother and the other children were not present. For the first time, Linus and Ava Helen learned that their son and his wife had separated and were to be divorced, and she was taking their five children with her to live in her native Switzerland.

The sad family news made the peace award festivities less happy for everyone. The Oslo ceremonies were a much simpler affair than those in Stockholm, with fewer social festivities and less protocol. In many ways, the second award was a let down for the Paulings. Many of the newspapers that had been so enthusiastic about the thirty-year-old Pauling who won the Langmuir Award were disgruntled when the sixty-two-year-old peace activist received the Nobel Peace Prize. *The Portland Oregonian* referred to him as a "controversial scientist from Pasadena." There were no letters to the editor mentioning that he had been born in Portland and that he was the first person on earth to receive two unshared Nobel Prizes. *Life* referred bitterly to the "Weird Insult from Norway," while the *New York Herald Tribune* thought that the award to Pauling "associates this semisacrosanct honor with the extravagant posturings of a placarding peacenik." The *New York Times,* on the other hand, commented, "While others have spoken for nations, ideologies, and above all, power, Linus Pauling spoke for Man. If the human race is to be spared the ultimate holocaust, it too will speak, and for a long time to come, of Linus Pauling."

Congratulatory messages streamed in from all over the world; from U Thant, Bertrand Russell, Albert Schweitzer, Pablo Casals; from Professor and Mrs. Kaoru Yasui, from Sean O'Casey, from his old friend, Barry Commoner, of Washington University in St. Louis, who wrote, "All scientists are in his debt."

When Pauling returned from Europe, after giving his Nobel address at the University of Oslo on December 11, he received a tremendous ovation from more than three thousand persons who packed the Grand Ballroom of the Commodore Hotel in New York City to pay him honor. M. S. Arnoni, of the newsletter *Minority of One,* had arranged the tribute to the "relentless crusader for world peace." In Los Angeles, the American Civil Liberties Union and the Women's Strike for Peace were arranging affairs to honor him, each of which was expected to attract audiences of over a thousand.

The historian Henry Steel Commager addressed the New York audience, giving credit to Pauling and other dissenters for creating the political climate that made it possible for President Kennedy to sign the partial test-ban treaty. The mood of the audience of three thousand was jubilant, and men and women who could not be seated stood throughout the affair. Others attended a dinner sponsored by the American Civil Liberties Union at the Beverly Hills Hilton Hotel. Spotlights picked out the figures of the Hollywood celebrities in attendance, as Linus and Ava Helen looked down at the crowd from the speakers' platform. Seated with them were Linus's two sisters, Lucile and Pauline. The three children of Belle and Herman Pauling were now older than either of their parents had lived to be, and it had been a long time since they had shared a meal together.

Pauling had threatened to sue Senator Dodd if he accused him of Communist activity outside the sanctuary of Congress. This may have seemed a rhetorical gesture. In the aftermath of the hearings, however, Pauling did file a number of lawsuits against newspapers and magazines that made similar accusations in print. Indeed, over the years Pauling was often quick to seek legal redress when he felt that he had been slandered or damaged in some way.

On October 10, 1960, the *St. Louis Globe-Democrat* published an editorial titled "Glorification of Defeat" that claimed that Pauling had refused to supply the committee with the names of the signatories of the petition and further stated, "Pauling contemptuously refused to testify and was cited for contempt of Congress." This had, in fact, never happened and the editor published a retraction.

Pauling was not satisfied with the retraction, since the paper went on to make further statements implying that he was lacking in patriotism, and using the phrases "deceit and evasion" and "Communist conspiracy" in connection with him. He took the paper to court, but the jury's verdict was in favor of the *Globe-Democrat.*

Pauling had several difficulties in these lawsuits. First, by filing suit he was, in effect, reversing roles and attempting to use the power of the state to silence his opponents. Many of his liberal friends believed that this was wrong; that speech should be responded to with speech, not with legal intimidation. The second problem was that there was some basis in fact of the allegations. Pauling was, in fact, involved in political activities that included Communist participants, including some sponsored by organizations dominated by Communists. Jurors could easily be persuaded that the reporting was true, or at least close enough to the truth to exclude deliberate distortion.

The final problem was proving that he had been damaged by the reporting. In his legal briefs, Pauling had to argue that he had suffered tangible losses that should be recompensed financially. In a statement about the *Globe-Democrat* editorial, Pauling claimed:

> The completely untrue statements made by the St. Louis Globe-Democrat . . . have without doubt greatly damaged my reputation and impaired my integrity, and have also, in my opinion, served to cause readers of the Globe-Democrat . . . to take action that will cause me serious financial damage.[10]

In fact, there was no real evidence of financial or other damage. As Judge Henry J. Friendly observed in dismissing a $500,000 libel suit Pauling filed against the *New York Daily News,* Pauling "became an outspoken advocate of causes . . . sympathetic to communism . . . thereby creating for himself a reputation that could not and did not suffer any damages by reason of the editorial complained of."[11] The problem, Pauling concluded, was that his character witnesses were too convincing. The juries were simply not convinced that he had been hurt by the newspaper criticisms.

He did succeed in recovering damages in some of his suits, however, through out of court settlements. The Bellingham Publishing Company of Bellingham, Washington, paid $16,000 to settle a $500,000 suit. In this case, Pauling had also filed suit against the authors of several letters to the editor, one of whom had called him a Communist, and another who had referred to the "nefarious, treasonable activities of LP." The Hearst Corporation also paid $17,500 to settle a $1,000,000 lawsuit.

In a letter to Peter Pauling, Pauling hoped that "these libel suits will force newspapers to be more careful about what they print about me."[12] And, indeed they may have. Several newspapers, including the *Buffalo Herald-Press* and the *Syracuse Herald-Journal,* and *Life* magazine published retractions or clarifications in response to his complaints or offered him a chance to respond in a letter to the editor.

Pauling's most important lawsuit was against the *National Review,* edited by the celebrated conservative pundit William F. Buckley. On July 17, 1962, the journal printed an editorial denouncing a number of prominent liberals as "collaborators" with Communism. The list included the sociologist C. Wright Mills, a prominent lawyer, and two Yale University professors as well as Pauling. The editorial claimed that, while these people might not be Communists in the legal sense, "the objective fact is that these persons . . . have given aid and comfort to the enemies of this country." While these people could not be convicted of treason, since the country was not technically at war, the editorial urged that they should be denied "public respect, honor and rewards" such as "chairs in some of the nation's most noted institutions of learning."

Pauling received a letter from one of the Yale professors, who was considering filing a libel suit. The professor recommended a lawyer, Michael Matar. William F. Buckley was just as proud and certain of his views as Pauling. He felt that there was no way he could back down from Pauling's suit. In a second editorial, published after the suit was filed, Buckley opined that other publishers "may have been too pusillanimous to fight back against what some will view as brazen attempts at intimidation of the free press by one of the nation's leading fellow travelers." He, on the other hand, suffered from no such cowardice. The battle was on.

Buckley's strategy in the lawsuit was twofold: to prove that the accusations against Pauling were true, and to show that he had suffered no damages from them. In a legal brief filed with the court, *National Review*'s attorneys included a list of ninety-three instances in which Pauling "has been engaged in . . . Soviet-serving activities and Communist-aiding fronts, which in varying degrees reflected the desires and policies of the Soviet Union." The list included items such as the following:

- member of the Progressive Citizens of America
- signer of a letter in defense of Communist Party leaders
- voucher for Dr. Sidney Weinbaum, charged with perjury and fraud in failing to disclose his Communist Party membership under an alias, while he served as a Physicist at California Institute of Technology
- signer of statements on behalf of eleven convicted Communist leaders, published in the "Daily Worker"
- signer of a petition to Attorney General on behalf of Hans Eisler, deported Communist
- initiator of National Committee to Repeal the McCarran Acts
- signer of a petition to Attorney General requesting he drop deportation proceedings against Hans Eisler because he was a great artist
- initial sponsor of American Peace Crusade—cited as subversive by the usual authorities
- member of the National Committee to Secure Justice for Morton Sobell
- signer of statement to end Korean War by letting Prime Minister Nehru mediate it
- signer of statement denouncing House Un-American Activities Committee, circulated by the Committee of One Thousand
- refusal to answer under oath before California Senate Interim Committee on Education—"Do you or did you ever pay dues in the Communist Party?"
- sponsor of Conference of Inquiry into the Facts of the Rosenberg Case[13]

The list was made up largely of perfectly legal protest actions, activities, and events that were public knowledge and that, in fact, Pauling had been eager to publicize. The legal point, however, was not to prosecute Pauling, but to defend *National Review* from Pauling's claim that it had falsely described him as a Communist sympathizer.

Of course, some or many of the statements on the list may have been erroneous. The legal brief gives no evidence for them, other than to say that the first forty-four were from the House Un-American Activities Committee. Even if all of the specific items were true, however, the list would still present a distorted picture of Pauling's political activities because of its selectivity. It left out a great many affiliations that were clearly outside the orbit of Communist-dominated or Communist-infiltrated organizations.

In response to *National Review*'s allegations, Pauling prepared a list of his memberships in organizations. He tried to make the list complete and stated that no organization had been intentionally omitted. The list provides a useful inventory of Pauling's political activities, some of which coincide with the *National Review* list, some of which correct it. It included the following items:

- I have never been a member of the American-Russian Institute of Southern California. . . . I attended a dinner given by the Institute, approximately 1947
- I was a member of the Science Committee of the National Council of American-Soviet Friendship . . . which was attempting to arrange an exchange of scientific literature . . . [and] was awarded a grant of $25,000 by the Rockefeller Foundation
- I signed a statement in behalf of the ten Hollywood movie writers and directors, called the Hollywood Ten
- I signed a statement opposing the Internal Security Act, called the McCarran Bill
- Since the fall of 1959, I have been a sponsor of the National Committee for a Sane Nuclear Policy
- I am [or was] a member of Union Now . . . Progressive Citizens of America . . . the Society for Social Responsibility in Science . . . the American Humanist Association . . . the World Association of Par-

liamentarians for World Government . . . the International Institute for the Study and Development of Human Relations . . . the Federation of American Scientists . . . the Federation of Atomic Scientists . . . the Board of Sponsors of the Bulletin of the Atomic Scientists . . . the International War Resisters . . . the Emergency Civil Liberties Committee.

When presented with the *National Review*'s list in court, Pauling readily confirmed many of the activities and forthrightly defended them: "I decided that it was a good activity that should help the cause of international understanding and world peace." There was a long discussion of whether or not he knew that certain organizations followed a "Communist line":

Q: *Professor Pauling, do you know what the Communist Party Line has been over a period of years on all issues that have been mentioned in this case?*

A: *No, I would say that on very few issues do I know what the Communist Party line has been over the last eighteen years.*

Furthermore, Pauling insisted he didn't know how to identify a Communist:

Q: *Is it still the fact that you are unable to tell Communists, Professor?*

A: *Well, I think so. As I look at the audience here I don't know which of these people are Communists.*

Q: *Do you know how to tell a fellow traveler?*

A: *I don't think I am expert enough in this field to answer this question.*[14]

Pauling's attorneys were hard put to prove personal malice of the *National Review* staff. No one on the staff knew him personally; their differences were clearly political. Nor could Pauling's attorneys show financial damages or damage to Pauling's reputation. In 1961, the year before the *National Review* editorial, Pauling was one of the scientists honored by *Time* magazine as "Men of the Year." Indeed, one of the main gripes in the *National Review* editorial was that people such as Pauling were being honored and respected by society, not ostracized as subversives.

In response to the pretrial examination, Pauling submitted data showing that his net income, after expenses, had increased from $40,403 to $49,984 from 1962 to 1964. His income from lectures and articles had increased from $8,218 to $12,260. In 1964, his textbook royalties of $29,394 exceeded his salary of $25,000. Despite *National Review*'s urging that Pauling and others be denied honor, employment, and other sources of income, he continued to do quite well.

After Pauling's side in the lawsuit presented its case, Justice Samuel Silverman dismissed the complaint without hearing from the defense or sending the case to the jury. In doing so, he set an important legal precedent. He ruled that people such as Pauling should be considered to be "public figures." This meant that libel suits against them should be held to a higher standard than lawsuits against ordinary citizens, the same standard applied to libel suits against public officials. Under this standard, it would be necessary to prove that the statements were made with "actual malice"—with knowledge that they were false or with reckless disregard of whether or not they were false.[15] Justice Silverman ruled that, under this standard, Pauling's case against *National Review* was not strong enough to merit continuing the trial.

Throughout his life, Pauling spoke his own mind, clearly, independently, and, when threatened, courageously. This is why the efforts of the Internal Security Committee and the *National Review* to brand him as a "fellow traveler" were doomed to failure. He simply did not fit that mold, either in his political views or affiliations or in his personality. He did not modify his views to fit changes in the Communist party's or any other group's political "line." While he sometimes agreed with the Soviet Union and the Communists, on other occasions he disagreed strongly with them. When the Soviet Union broke the moratorium on nuclear testing, he lodged a strong public protest. His strong opposition to Soviet science policy, particularly with regard to theories of chemical bonding, was well known.

When called up before the committee, Pauling planned his own defense with none of the tactical hesitation and use of the Fifth Amendment that sometimes characterized the behavior of witnesses who were actually following party discipline. He felt confident in suing *National Review* and other publications because he knew they were wrong in branding him a Communist.

On the other hand, if one assumes the mind-set of the committee members or the *National Review,* one can see how they might have been confused. Pauling's views on many issues were strongly anti-American. He advocated friendship and reconciliation with the Soviet Union at a time when the Soviets were, in fact, an authoritarian dictatorship and threatening military power.

Pauling's political views during this period were those of a New Left radical, very much in tune with the positions expressed by groups such as Students for a Democratic Society. He opposed the Vietnam War and believed America was an imperialist, exploitative nation, responsible for much of the poverty and suffering in third world countries. He wanted to redistribute the world's wealth to give more to the poor. And he saw the American power elite as the chief roadblock to necessary change.

The clearest statement of Pauling's political philosophy is the handwritten text of a speech he gave in 1968 to a seminar dinner in Los Angeles on the topic "What Can a Man Do?" The dinner, with informal dress and a prix fixe of $12.50, was sponsored by the Center for the Study of Democratic Institutions. Portions of the speech were reprinted in *Center Magazine,* as part of a colloquy with other resident thinkers.[16]

In many ways, the speech sounds as if it could have been given by Upton Sinclair if he had survived into the 1960s. In the best of Sinclair's radical tradition, Pauling painted a picture of struggle between the forces of good and evil. The handwritten text shows the strength of his angry feelings, with key words capitalized or underlined. Good was represented by the common people of the world, evil by "the *rich,* the *powerful,* the *establishment.*" Pauling thundered, "*I believe in non-violence.* But the *Establishment* believes in *violence,* in force, in MACE, NAPALM, Police power, aerial bombing, nuclear weapons, war."

He observed, "I always feel more certain about my understanding of a problem if I have some figures," and went on to quote figures about the unjust distribution of wealth in the world, with wealthy Americans having three hundred times the income of poor third world people. The rich would never voluntarily give up their wealth, so the people must rise up and take it from them:

> I believe that it is only through continued and vigorous protest by large groups of people, through demonstrations by the mass of the people, through revolution, that the evil future for the world that I foresee as a possibility can be averted.

Pauling went on to denounce the CIA as "the most immoral institution in the world," because of its role in suppressing revolutions in Indonesia and Guatemala. Because of the perfidy and violent inclinations of the powerful, Pauling argued,

> As long as the selfishness of the Establishment remains the determining factor, our hope that the coming revolution will be *non-violent* has little basis in reality.

Like all ideological positions, Pauling's New Left views had their weak points. He was clear as to what he was against, but vague as to what alternative he proposed. He was quick to point out the flaws in America's democratic institutions but reluctant to consider the actual record of those countries that have had violent revolutions to correct similar flaws. One of the discussants at the Center for the Study of Democratic Institutions observed at the time:

> Dr. Pauling indicates that he would tear up the Constitution, because that is what revolution means, so that the republic that we know will no longer exist in his world. In Indonesia, where he said untold thousands were murdered by the C.I.A. because it was afraid of democracy, what kind of democracy was he talking about? Sukarno was the leader of what one might refer to as Dr. Pauling's democracy. . . . Dr. Pauling's revolution, since it is a revolution, must produce some kind of government, but he didn't tell us what it would be.[17]

In response to this criticism, Pauling backtracked a bit, acknowledging that the United States was a comparatively free country, and stating that he would like to continue with American constitutional provisions if possible. However, he opined that the United States was not free if young men could be drafted into an immoral war in Vietnam. He wanted government financing of political campaigns, government ownership of television stations, and "greater governmental control over monopolistic industries, over big corporations." Only by making these changes voluntarily could we avoid the "violent revolution that I fear is going to come."

There is a striking contrast between Pauling's scientific thinking, which was innovative and highly complex, and his political thought, which was simple and predictable. His political speeches were similar to those being given by thousands of other New Left radicals at the time. Even his examples and illustrations were constantly used in the rhetoric of the time. In his scientific work, he sought out difficult unresolved problems. In his political rhetoric, he avoided the difficult issues, such as how to reconcile revolutionary egalitarianism with the need for economic incentives and human rights. His views were sincere and well motivated, and he had every right to expound them freely. But they did not draw on the sophistication and creative insight that characterized him as a scientific thinker.

8

Orthomolecular Medicine, 1966–1990

Before going to Oslo to receive the Nobel Peace Prize in 1963, Pauling had asked for and accepted the invitation of his old friend, Robert M. Hutchins, president of the Center for the Study of Democratic Institutions, to go to Santa Barbara as a staff member. This was in some ways a surprising decision since the institute had no laboratory facilities and Pauling had shown no signs of being ready to retire from active involvement in chemical research.

Ava Helen explained that Santa Barbara had its attractions:

> One of the reasons which makes it so attractive to us is that we are that much nearer our ranch where we hope to spend a good bit of time. We are now busy getting an architect and in odd moments hope to make a start on our house. Also Santa Barbara is a lovely place and this appointment gives Linus complete freedom; he can do whatever he wishes, work on medical problems, write books, stay where he wants and in addition is a 25% increase in salary.[1]

In many ways, however, Pauling seemed more eager to leave Caltech than to commit himself to the work of the institute. Caltech was increasingly unfriendly to Pauling as he became more prominent as an activist and less active as a physical chemist. His valence bond theory became less important in chemistry as the field emphasized molecular-orbital and other approaches, and the coolness of many people at Caltech contrasted sharply with the enthusiastic admiration he received from people in the peace movement.

In 1958, Pauling had resigned as chairman of the Division of Chemistry and Chemical Engineering, after twenty-two years. Although this resignation was formally voluntary, to allow more time for his teaching and research, in fact he was pressured into resigning by President Lee DuBridge, who was under pressure from many members of the board of trustees to censure Pauling on political grounds. Indeed, Pauling says the president told him that he would have been dismissed from the university if he had not had tenure.

Pauling's base of support within the Division of Chemistry had also eroded, in part because some colleagues did not approve of his politics, but more importantly because many felt he was spending too much time traveling around the world giving speeches and too little doing administrative and scientific work at home.

Despite the pressures, Caltech respected Pauling's academic freedom throughout the McCarthyist period, and he remained a tenured professor with complete freedom to pursue his research. Laboratory space, however, was at a real premium at Caltech, and Pauling was no longer in a position to command as much as he felt he deserved. He had started a number of projects that involved searching for biochemical explanations for mental illnesses in the bodily fluids of mental patients. These projects were not as successful as he hoped, and when it was time to reallocate lab space, some members of the division suggested that he could easily cut back by abandoning his mental health studies. They also believed that some of the researchers Pauling had hired with his grant money did not have the qualifications usually expected of Caltech faculty. One was a physician who had been fired from a job at a hospital because of his refusal to sign a loyalty oath. This man was dedicated to the mental health project but had no experience as a researcher.

Pauling's colleagues also felt that he was too busy traveling or attending to his own work to provide adequate supervision to the younger researchers he had recruited. One of Pauling's close friends at the time says that Pauling considered himself the emperor of chemistry, yet he was not respected and couldn't get lab space at his own institution.

The event that triggered Pauling's departure from Caltech was the deep hurt he experienced when the Chemistry Division took no official notice of his Peace Prize, not even so much as a newspaper clipping on the bulletin board. The Biology Division gave him a party, but the Chemistry Division did not. Many of the professors believed that Pauling's peace advocacy had nothing to do with chemistry and attracted unwanted notoriety to the department. Some of the physicists at Caltech had worked on assembling the first atomic bombs and took Pauling's assaults on nuclear testing as a personal affront. When asked about the award, President Lee DuBridge said that he preferred to work for peace in ways that were not recognized by the Norwegian Committee. Although some individuals were supportive, the predominant mood on campus seemed to reflect *Life* magazine's view of the prize as a "weird insult from Norway."

Ava Helen thought that the lack of recognition for the Peace Prize was the principal reason Pauling decided to leave Caltech, despite his fondness for the school, which was both his graduate alma mater and his principal employer for his entire professional career. He did agree to accept a nominal unpaid appointment as a "research associate" on leave. He continued to be listed as a Professor Emeritus at Caltech after he reached the age of seventy, and the school made up for any slights of the past by giving him an enormous party on the occasion of his ninetieth birthday. Five Nobel Prize winners spoke, and Pauling impressed everyone by giving a rousing scientific speech of his own.

Although they had lived in Pasadena for forty-one years, the Paulings were world citizens and had acquaintances and supporters in many places where they might choose to settle. At sixty-three, Pauling was at an age when many men think of retirement, and Santa Barbara was closer to the Ranch. He and Ava Helen had built a new country house on their Big Sur property, a house roomy enough to hold a small conference on world peace, or their fifteen grandchildren if all of them visited at once. Pauling, following tradition, laid out the original plans for the complex structure himself and acted as general contractor.

The low redwood and stone structure, tinted a grayish green to match the ocean it overlooks, had three wings. Pauling laid it out to fit the irregular terrain and to suit the family's needs. The bedroom-study wing was built on the ocean side on the top of a rocky knoll, a hundred feet above mean high tide. There were two studies, one for the professor, one for his wife. The massive fireplace, built of native stone, was in this study wing.

The living room–kitchen–dining room area was at an intermediate level in a second wing on the Salmon Cove side. Set at an angle to the other wings was the third wing with its guest bedrooms, its garage, its library stack room. The garage was also lined with bookshelves.

The inside walls of the new Ranch House were of local redwood and rock. Great glass windows let in the sun and shut out the fog. The roof lines were sloped like a sea gull's wings dropping earthward. In the center of the knoll was a partly enclosed patio that was dominated by a single native boulder, six feet around.

Their Pasadena home remained in the family. It was purchased by their daughter and son-in-law, who had outgrown their smaller home in Altadena. Linus and Ava Helen found a satisfactory small house in Montecito, a few minutes from the center, and moved their favorite items of furniture there: the bright Oriental rugs, Scandinavian chairs, and decorative objects from abroad and the long table on which they once spread out petitions received from scientists all over the world. Santa Barbara itself is a lovely resort community with a pleasant, smog-free climate, beaches, and mountains.

The Center for the Study of Democratic Institutions was a liberal think tank. The permanent staff, consisting of thirty-five to forty persons, included a variety of scholars from different disciplines: economics, history, mathematics, philosophy, theology, political administration, sociology, and journalism. Pauling, like the rest of the staff at the center, had no specified duties other than attending, when in Santa Barbara, the discussions held each weekday morning around the great tables in the large room off the terrace. All the conversations were tape-recorded, and many of the tapes were made available to radio stations, universities, and study clubs. Among these was a conversation between Linus Pauling and Helmet Krauch on "Science: For Truth or Good," in which Pauling took the position that scientists should be perfectly free to follow the lead of their curiosity wherever it takes them.

Pauling was the only physical scientist at the center when he was there and had no laboratory facilities. He could do theoretical writing on scientific issues, but there was no one who could understand his basic scientific work. Nor had Pauling ever focused his serious creative efforts on social or political topics. In fact, ever since his high school days he had avoided the study of history or the social sciences.

The diversity of training and vocabulary of the fellows at the center sometimes made communication difficult. At times the conversation was dull and uninspired; at others the cross-fertilization could lead to exceptional insights. Pauling was eager to talk about social issues such as peace, racial injustice, and civil liberties but was bored by lengthy discussions of the United States Constitution. When he talked, he tended to set the pace and dominate the conversation. The others accepted his doing so up to a point, because he was always interesting, but eventually the discussion had to return to social science issues. After avoiding social science all his life, Pauling found himself surrounded by it, and he seemed restless.

Ava Helen described their disenchantment with the center in a letter to a friend:

> I am not at all sure that this situation at the Center is going to be satisfactory because I am afraid that after a short time, Linus is going to decide that it is altogether too superficial. Hutchins is a brilliant and witty person, but I think is rather superficial himself. He gives the impression of thinking that almost anything can be solved if one is just witty enough. Perhaps I am too pessimistic. Anyway, we shall see.[2]

Pauling, at a less than diplomatic moment, was said to have referred to Hutchins as "a windbag."

The Paulings had many friendly acquaintances at and around the center, but they generally kept to themselves rather than forming close friendships. One informant thought that Pauling tended to dominate Ava Helen but was also very dependent on her. The four years the

Paulings spent at the center were a time of recuperation. No single social or scientific issue sparked great excitement in Pauling. The arms race continued, and nuclear testing was accelerated in underground tests, but public concern had subsided since the atmospheric test-ban treaty had ended the fallout problem. The Paulings traveled frequently, and Linus spoke at all sorts of functions.

It was at this point in his life, at the age of sixty-five, that Pauling traveled to New York City to give a speech at the Carl Neuberg Medal award dinner. He casually mentioned that he hoped to live for another fifteen or twenty years in order to observe new developments in science and society. A few days later he received a letter from a biochemist, Irwin Stone, who had been in the audience. Stone promised him the chance to live for another fifty years if he would take massive doses of vitamin C. Perhaps because he was not absorbed with anything else particularly pressing, Pauling decided to take this suggestion seriously. At the very least, it was a scientific question, and one that could have significant social consequences as well.

Thus began Linus Pauling's last great crusade, one that was to be every bit as controversial as his crusade against nuclear fallout. Linus and Ava Helen followed up on Stone's suggestion, became true believers in his theories, and threw the full weight of Linus's scientific reputation behind them. A central focus of his life in the decades after 1966 was to be his campaign for what he called "orthomolecular medicine": the maintenance of health and cure of disease by regulating the concentration in the body of substances naturally found there.

In joining this crusade, Pauling associated himself once again with an active social movement. The use of natural foods and remedies is an important aspect of the countercultural movements that grew up in the 1960s. Health food stores throughout the country flourish by selling "natural" products free of preservatives, artificial sweeteners or flavorings, or refined sugars. A significant body of literature already existed to support their claims, but little of it was written by established scientists. Certainly no scientist of Pauling's stature had come forth to defend the claims of the health food movement.

Recognized professional nutritionists were skeptical of many of the claims of health food enthusiasts. This, however, did not discourage

Pauling, who had opposed official opinion before and been vindicated. He suspected that this was an issue in which the powers that be, such as the American Medical Association, the drug companies, and the food processors, had suppressed a basic truth.

Although Pauling had actually never done any research on vitamins or nutrition, the topic of vitamins fit in with his interest in molecular structure. He had worked on other medical topics and certainly had many of the skills needed to do medical research. He set about reading the literature on vitamin C. Most of the publications were not new, nor did the topic generate much interest among medical or dietary researchers. The scientists who had done the studies had largely gone on to other topics. Pauling, however, reviewed their findings and reached conclusions that were startlingly different from those of most of the scientists who had done the actual research.

Vitamin C was first isolated and analyzed by the biochemists Albert Szent-Gyorgyi, Charles King, and others in the years 1928–33. They found it to be present in large amounts in Hungarian red peppers, as well as in citrus fruits, tomatoes, berries, fresh green vegetables, and potatoes. Szent-Gyorgyi found the chemical formula for the substance to be $C_6H_8O_6$ and gave samples to an English chemist, W. M. Haworth, who determined its molecular structure. They then named it ascorbic acid, meaning the acid that prevents the disease scurvy.

The effectiveness of citrus fruits and fresh vegetables in preventing scurvy was known as far back as the eighteenth century. Scurvy is a deficiency disease, the symptoms of which include weakening of the capillaries, hemorrhaging, bleeding of the gums, loss of teeth, anemia, general debility, lung and kidney disorders, and, within a few months of ascorbic acid deprivation, death. While it is uncommon among people who have access to an adequate diet, it was a major problem in the past when fresh fruits and vegetables were unavailable in the winter in northerly climates. It was especially acute among sailors who spent long months at sea eating dried and preserved foods, as well as in poorly fed institutionalized populations and in communities suffering from famine. More than half of the crew died of scurvy on Vasco da Gama's trip around the Cape of Good Hope in 1497–99, and massive deaths from scurvy were not uncommon during long sea voyages in the period of European expansion around the globe.

Gradually, people began to make the connection between lack of a proper diet and scurvy. Finally, in 1747, the Scottish physician James Lind carried out an experiment in which he treated some scurvy patients with citrus fruits, and some with other foods or medicines. The recovery by the patients given fresh fruit was so dramatic that he could justifiably claim to have found the cure for scurvy. Unfortunately, acceptance of this finding was slow and it was not until 1795 that British sailors were routinely given their daily ration of fresh lime juice, and many more years until the same was given to all merchant seamen.

While the dietary cure for scurvy has thus been known for over two hundred years, it was not until the early twentieth century that researchers began to identify in many foods small amounts of substances now known as vitamins, needed to prevent diseases such as scurvy, beriberi, pellagra, and rickets. Only when ascorbic acid was isolated chemically was it possible to study its precise role in human biochemical reactions, and some of its functions still may not be understood. Ascorbic acid is known to serve as a necessary coenzyme in the synthesis of collagen molecules. Collagen, essential to the body's connective tissues, is composed of protein molecules stretched out in long elastic fibers. Collagen disorders are associated with diseases such as rheumatic fever and rheumatoid arthritis as well as certain skin disorders, and of course scurvy.

The importance of vitamin C in preventing scurvy was quickly recognized in the 1930s, and the ingestion of a small amount of the substance, about 10 to 50 milligrams a day, is universally accepted as essential to human nutrition. Of course, any balanced diet will include more than this amount. With this basic discovery, most scientific interest in vitamin C ended. A few scattered researchers, however, continued to study the effects of vitamin C in other illnesses. In some of these studies, doses very much larger than those required to prevent scurvy were used—doses often ranging from 1 gram (1,000 milligrams) a day, up to 10 or even 25 grams. While these doses seem very large in comparison to the 45 milligrams recommended for adults by the U.S. government, they amount to no more than a teaspoonful or two of pure ascorbic acid a day. Other studies used relatively smaller doses, often 200 milligrams.

At this point, Linus Pauling's contribution to the debate on vitamin C was based entirely on his interpretation and popularization of the results of these previously published studies. In interpreting this literature, Pauling consistently emphasized positive results obtained with the use of vitamin C, while explaining weak or negative results as resulting from deficiencies in research design. This approach is the opposite of that of most of the medical and scientific establishment, which generally concluded that there was insufficient evidence to justify using vitamin C as a remedy or preventative for the common cold.

In 1967, Pauling left the Center for the Study of Democratic Institutions to accept an appointment at the University of California in San Diego. This gave him the opportunity to work in the laboratory again and put him in contact with other research chemists. Among them was Arthur B. Robinson, a young professor in the department, who became interested in Pauling's work and began to collaborate with him. Robinson, born in Chicago in 1942, first met Pauling when he was a student in Pauling's freshman chemistry class at Caltech in 1959. He was enthralled with Pauling's lectures. One of his friends once remarked to him, "I don't know which is more fun, watching Pauling lecture or watching you watch Pauling lecture."

After he had completed his Ph.D. at the University of California at San Diego, Robinson was considered to have such great potential that he had the unusual honor of being invited to join the faculty of his own graduate institution. His early research was on peptides, protein synthesis, and the molecular timers of protein turnover and aging. Robinson was excited by the opportunity to collaborate with Pauling.

Senior members of the department were unhappy about Pauling's preoccupation with orthomolecular medicine, however, and he alienated more conservative members of the administration by his defense of Herbert Marcuse, a prominent Marxist philosopher who was being attacked by anti-Communists.

In 1969, Pauling moved to Stanford University, which was closer to his ranch and promised him laboratory facilities as well. Robinson was eager to follow Pauling to Stanford, but the department at San Diego had given him two years free of teaching with the condition that he would teach the following year. He met his obligation to San Diego by commuting between the two campuses, teaching courses at San Diego and running Pauling's laboratories at Stanford.

In 1970, Pauling published a popular book, *Vitamin C and the Common Cold.* In it, he claimed that most colds could be prevented by regularly taking large doses of vitamin C, and that massive doses of the vitamin could quickly cure a cold that had already started. In the introduction, Pauling stated that he and Ava Helen had begun following Irwin Stone's high vitamin C regimen in 1966, and that they had noticed "an increased feeling of well-being, and especially a striking decrease in the number of colds."[3]

It may be, however, that this decrease in cold symptoms was not as striking as Pauling implied. In a letter to his son Peter on January 9, 1969, Pauling wrote, "Both mama and I have suffered from colds during the last few weeks, never very bad but dragging on. I don't think that we had the flu, because at no time did we have a fever."[4] Over the years there have been occasional press reports of rumors that Pauling was suffering from a cold. In interviews, he admitted to having "allergies" but claimed that he was able to ward off any threat of a cold by increasing his dose of vitamin C.

Pauling did not let his and Ava Helen's dragging colds in the winter of 1969 interfere with the success of his book. *Vitamin C and the Common Cold* received tremendous publicity and drugstores throughout the country had their stocks of vitamin C cleaned out by an eager public. The book was reissued in paperback in 1971 and 1973, and an expanded version, *Vitamin C, the Common Cold and the Flu,* promised to ward off the expected swine-flu epidemic in 1976. It was followed by *Cancer and Vitamin C,* coauthored with Ewan Cameron in 1979 and *How to Live Longer and Feel Better* in 1986. All were well received by the public.

Pauling's books on vitamin C accused the medical and drug industry establishments of hushing up its effectiveness either because of ignorance or because of their interest in marketing more expensive preparations that offer only limited symptomatic relief.

What was the evidence for these remarkable assertions? Research into the use of vitamin C for the common cold began as early as 1936 in Germany and has continued sporadically since. Most of the studies compared the frequency and severity of cold symptoms experienced by an experimental group receiving ascorbic acid with those of a control group not receiving supplementary ascorbic acid. In most cases, small differences were found between the two groups.

The study Pauling attacked most vehemently and for the longest time was the decade-long study "Vitamins for the Prevention of Colds" published by Cowan, Diehl, and Baker, all of the University of Minnesota. Dr. Donald Cowan was director of the Student Health Service and of the School of Public Health and had also done research on allergies and pollution. Dr. Harold S. Diehl was dean of medical sciences, and Dr. Abe B. Baker was professor of neurology and had done research on infections of the nervous system and on arteriosclerosis. To their disappointment they concluded that their research on cold prevention had no important results to offer cold sufferers.

When Pauling, in 1970, discovered the 1942 study, he wrote a critical letter to the *New York Times*, claiming that the authors had not adequately reported their positive results. Dr. Diehl replied in a letter published on December 26, 1970:

> In his December 15 letter, Prof. Linus Pauling referred to several "controlled studies" of the use of vitamin C for colds which gave "positive" results. He then quotes from one of these studies published in the *Journal of the American Medical Association,* Dec. 14, 1942, of which I was one of three authors.
>
> He states, correctly, that "these three physicians reported a 15 percent smaller incidence of colds in the students (at the University of Minnesota) who received 200 milligrams a day of vitamin C (throughout the winter months) than in those who received a placebo. The average difference between the two groups amounts to one-third of a cold per person."
>
> Professor Pauling, however, fails to mention the summary of this report which states: "This controlled study yields no indication that either large doses of vitamin C or large doses of vitamins A, B_1, B_2, C, D and nicotinic acid have any important effect on the number or severity of infections of the upper respiratory tract when administered to young adults who presumably are already on a reasonably adequate diet."
>
> Professor Pauling also fails to mention a report, by two of us who were

authors of the paper referred to above, on the effectiveness of vitamin C and antihistaminic agents in the early treatment of the common cold—*Journal of the American Medical Association,* Jan. 3, 1950.

The summary of this report states: "Under the conditions of this controlled study, in which 980 colds were treated in 367 supposedly non-allergic students, there is no indication that ascorbic acid (vitamin C) alone, an antihistamine alone, or ascorbic acid plus an antihistamine have any important effect on the duration or severity of infections of the upper respiratory tract."

This exchange typifies the differences between Pauling and the researchers who had conducted controlled clinical studies of vitamin C and the common cold. If the researchers found differences between the experimental and control groups, they were small. Many studies found no difference. The researchers did not believe that there was sufficient basis for recommending large doses of vitamin C.

Pauling interpreted the results differently and thought that the researchers were too conservative in reporting their results. This kind of disagreement about the interpretation of scientific data is legitimate and can provide the impetus for additional research that may overthrow the conventional wisdom in the field. It is possible that new research, such as that being conducted by Professor Elliott Dick at the University of Wisconsin in Madison, will provide a firmer scientific basis for Pauling's conclusions. Pauling, however, did not believe that additional research was necessary before informing the general public of his conclusions. He was positive that his interpretation was correct and suspected that the researchers who disagreed with him were biased because of their connections with drug companies or physicians who had a vested interest in treating colds with other remedies. Without waiting for more research, he took his case to the public with popular books, lectures, and interviews with the mass media.

The established medical journals and authorities responded to the increased public interest in vitamin C in a predictably conservative manner. The *American Journal of Medicine* and the *Journal of the American Medical Association* both published reviews of the literature that concluded that the evidence was not strong enough to justify using vitamin C as a preventative or treatment for the common cold. The *British*

Medical Journal, after reviewing both the American and Canadian studies and two British studies, which showed no significant differences, published an editorial commenting, "Major advances in treatment are usually apparent after a few well-conducted studies, and at present no strong evidence can be found to support the routine prophylactic use of ascorbic acid in well-nourished people."

These conclusions no doubt satisfied the large majority of physicians and scientists, who had more important things to do than to worry about a controversial treatment for a minor disease that is self-limiting in any event. Pauling, however, was deeply disappointed. He seemed to take the rejection of his message as a personal affront, and he attributed questionable motives to those who disagreed with him. As he observed in 1986:

> Fifteen years ago I was writing *Vitamin C and the Common Cold.* I was pleased with myself. I had made many discoveries in chemistry and other fields of science and had even made some contributions in medicine, although it was not clear that these contributions would have much effect in decreasing the amount of suffering caused by disease. Now, I thought, I have learned about something that can decrease somewhat the amount of suffering for tens of millions or even hundreds of millions of people, something that had been noticed by other scientists and by some physicians but for some reason had been ignored.
>
> I thought that all I needed to do was to present the facts in a simple, straightforward, and logical way in order that physicians and people generally would accept them. I was right, in this expectation, about the people but wrong about the physicians, or perhaps not about the physicians as individuals but about organized medicine.[5]

The more the establishment ignored and deprecated his arguments, the more vigorous Pauling became in asserting them. In his speeches and popular articles, he greatly expanded his claims. He did not hesitate to say that he believed that vitamin C and megavitamins, in the amounts he took and advised others to take, could cure or alleviate not only the common cold and cancer, but mental illness, viral pneumonia, hepatitis, poliomyelitis, tuberculosis, measles, mumps, chicken pox, viral orchitis, viral meningitis, shingles, fever blisters, cold sores, canker sores, and warts. These suggestions were not backed up by clinical studies but were based on theories he had developed to explain *why* vitamin C was so potent in helping the body to resist disease.

Pauling's greatest scientific contributions have always been as a theorist, synthesizing and extrapolating from data gathered by others. As a young chemist, of course, he also did laboratory research. As he got older, he became increasingly involved in administrative work and in supervision of the work of others, as well as in purely theoretical writing. He applied the same approach in his work on megavitamins; instead of doing empirical research himself, he sought to integrate and explain facts that he thought had been established by others.

At first, he turned to the theory of evolution in an attempt to explain why human beings are unable to synthesize vitamin C in the way that almost all other animal species do. Only the guinea pig, an Indian fruit-eating bat, and a few species of birds, in addition to primates, are known to require ascorbic acid in their diet rather than synthesizing it themselves. Pauling argued that these species lived in an environment where a very large amount of ascorbic acid was present in readily available foods. When a mutation occurred, as a result of radiation or some other chance factor, causing individuals to be born who were unable to synthesize ascorbic acid, these defective individuals suffered no ill effects because they were taking in sufficient quantities in their food. They also had a competitive advantage in that their bodies were relieved of the effort needed to synthesize ascorbic acid. Gradually, these individuals came to predominate in the human species, and among a few other species in restricted environmental circumstances.

While it is not unlikely for a mutation to cause an individual to lose the capability to synthesize an enzyme, it is very unlikely for a reverse mutation to cause the ability to be regained. Losing an ability requires only damaging the relevant genes; replacing it would require somehow re-creating them. (Although it is possible for a gene to be turned off and then turned on again.) Thus, in Pauling's theory, when human populations moved to colder climates where fresh fruits and vegetables were not so readily available, they did not recover the ability to synthesize ascorbic acid although they may have developed increased capacities to retain small amounts of vitamin C in the body rather than losing it in the urine.

University of California scientists Thomas Jukes and Jack Lester King have criticized Pauling's arguments on the grounds that the loss of the ability to synthesize vitamin C would not be adaptive if the need for it were as great as Pauling claimed. They also disagree with his estimate of the amount of vitamin C ingested by primitive humans, noting that many of the sources he cited were cultivated plants that were in short supply before the invention of agriculture. They note that many species with diets high in vitamin C have nevertheless retained the ability to synthesize it, and they postulate that the loss of the ability to synthesize vitamin C by humans and some other species was probably a neutral evolutionary change that occurred sporadically by mutation.[6]

Pauling argued that if human beings were to rely on a diet rich in raw vegetables with a high content of vitamin C, their intake of ascorbic acid would be several grams a day, much greater than the 45 milligrams recommended by the government for a male adult. Since contemporary diets are richer in meats, fats, and other foods with less vitamin C than certain vegetable diets would provide, he recommended that most people take 1,000 milligrams a day, or more, in order to maintain optimum health. He argued that by combining good dietary practices with proper doses of vitamins, low sugar intake, and no cigarette smoking, people could probably extend their healthy life span by approximately twenty years.

Of course, this phenomenal increase in health would not result merely from control of the common cold. In fact, no one had suggested that vitamin C was a specific treatment for cold viruses. Rather, the theory advanced by Pauling, Irwin Stone, and others was that the vitamin contributes to the body's ability to resist infections in general. Several different theories were advanced to explain this alleged almost panacean effectiveness of vitamin C. One theory is that the well-known importance of vitamin C in the synthesis of collagen contributes to health by strengthening the connective tissues. This means that burns and fractures heal more quickly, some sorts of back troubles may be cured or prevented, and the ability of body cells to resist or contain the spread of malignant tumors may be increased. The fact that collagen is depleted in scurvy, however, does not mean that giving large doses of vitamin C to an individual without scurvy will lead to the production of large amounts of collagen. Another theory was that vitamin C has antiviral and antibacterial effects by increasing the ability of cells in the blood to absorb and destroy disease-causing organisms.

It was also suggested that patients receiving large doses of vitamin C were unlikely to contract viral hepatitis after blood transfusions, and researchers suggested that maintaining a large concentration of vitamin C in the blood might increase resistance to any number of bacterial or viral infections. Furthermore, vitamin C was claimed to reduce the levels of cholesterol in the blood perhaps by increasing the rate of its conversion of bile acids. Even if excess vitamin C was excreted in the urine, this might help prevent infections of the urinary tract. Pauling saw merit in all of these theories, so long as they reinforced his advocacy of vitamin C.

Some of these theories were supported by suggestive, exploratory research findings and/or clinical reports. None of them had been verified by rigorous, well-controlled experimental studies. Pauling, however, was convinced that the effectiveness of vitamin C in curing the common cold had been substantially proved, and he was confident that his other ideas about vitamin C would be verified as well. He continued to focus his energies on further advocacy and research on the role of ascorbic acid and other natural substances in human health.

This new focus of Pauling's work was not viewed with much enthusiasm by the Stanford University Chemistry Department or the administration. Having a Nobel Prize winning chemist on one's faculty is all well and good, but how much in the way of laboratory space and resources can a major university devote to his eccentricities? Pauling also supported leftist causes on campus at a time of considerable campus turmoil. When a tenured professor of English, Bruce Franklin, was fired for participating in Maoist demonstrations that disrupted the campus, Pauling picketed on his behalf. In 1973, when some of Pauling's grant funding ran out, Stanford persuaded him to become an unpaid professor emeritus without laboratory space. He was, after all, well past the usual retirement age.

Lacking the facilities he needed at Stanford to continue his research, Pauling decided to follow up on a suggestion that Arthur Robinson had made to him a few years before: that they start their own independent research laboratory. Robinson, eager to continue his close collaboration with Pauling, gave up his job in San Diego and moved to Palo Alto to join him.

The institute, known at first as the Institute for Orthomolecular Medicine, would seek support from foundations, government funding sources, and private donors. They had very little initial funding from grants they were able to take with them, however, and Pauling and Robinson had to support the institute personally. Pauling was able to do this from his savings; Robinson used an inheritance he had received. (Robinson says that Pauling never actually delivered the money he promised, telling Robinson that Ava Helen objected.)

Research in any medical field is much slower than the chemical experiments to which Pauling was accustomed. Even with the use of animals, it is often necessary to wait several years to conclude a meaningful experiment. Furthermore, the idea of vitamin C as a cure for the common cold was old news at this point. Millions of people were already using vitamin C for this purpose, although the evidence for its effectiveness was still controversial. Pauling needed some new ideas, and if at all possible some new evidence, if he were to maintain public interest and excitement in the new institute.

He believed that these new ideas were there for the asking in the existing literature. He had read the claims of a small group of dissident psychiatrists who were using nicotinic acid as a treatment of schizophrenia. This fitted into his general perspective of emphasizing the use of vitamins and other natural substances in curing illnesses, and he was eager to believe that their claims were justified. Those claims, however, were very difficult to evaluate. Schizophrenia, more than the common cold and much more than cancer, is difficult to study with an experimental approach. It has afflicted a large but undetermined percentage of the people of the world for many years. It does not follow a predictable hereditary pattern, nor is it the result of any identified physical weakness. Some patients have only one episode; others are ill for many decades. While the average length of hospitalization for a particular incident averages four weeks in the United States, many patients recover for a brief period and then return to the schizophrenic state. It is difficult to know when a patient has been cured.

Schizophrenia strikes most often between the ages of eighteen and twenty-five. The plight of these young people, many of whom are intelligent and physically strong, has stimulated a tremendous amount of research by geneticists, biochemists, clinical psychologists, psychiatrists, social workers, sociologists, surgeons, nutritionists, demographers, and others. So far no definitive cause or cure of what is called schizophrenia has been found, although promising findings sometimes create a temporary flurry of excitement in the media.

High expectations for a chemical cure have been followed by disappointments. Administration of thyroid extract was a popular treatment at the turn of the century. Dozens of papers were then written of deficiency of carbohydrate metabolism as the cause of the illness. For a time chemists were excited about the role of serotonin in the biochemical processes of the human organism and physicians tried it with enthusiasm as they have other short-lived chemical approaches.

Someone once noticed the resemblance of hallucinations caused by lysergic acid diethylamide (LSD) to those of schizophrenics, and some volunteers tried taking LSD in the hope that it would give them insights into the schizophrenic state. Unfortunately, some recovered poorly from the experience and no useful insights were recorded. Others have observed that epileptics are not subject to schizophrenia, and shock treatment was developed in an attempt to introduce artificial epileptic seizures. For a time, lobotomies were used to remove part of the brain in an attempt to control the problem. All sorts of blood and urine studies show differences between schizophrenics and nonschizophrenics, but none of the differences had led to a cure. Psychiatric medications and psychotherapy can be effective in helping many patients to control their symptoms and live outside the mental hospital, but they do not offer a cure for the underlying disorder, which most specialists believe to be chemical in origin.

All sorts of treatments of schizophrenia have been tried from time to time, with apparent initial success in selected cases, but have failed to demonstrate lasting effectiveness. With little familiarity with this past experience, Pauling displayed remarkably little caution in rushing to the conclusion that nicotinic acid was a cure for schizophrenia. In 1973, he and David Hawkins edited a thick volume of papers that was published under the title *Orthomolecular Psychiatry: Treatment of Schizophrenia.* The book is aimed at a professional audience and contains articles of various types and quality. Some are rather detailed accounts of studies of the concentration and metabolism of various chemicals in the body. One of these is a study of the extent to which schizophrenic patients seem to absorb more vitamins than nonschizophrenics, which is interpreted as showing that they have a deficiency. This paper, coauthored by Linus Pauling with Arthur Robinson and six other people, made no attempt to assess the effect of megavitamin therapy on the patients' illness.

An article by Theodore Robie promised to review ten years' experience in using niacin in treating schizophrenia, although in fact it mentioned only four case histories in which treatment was successful. None of the articles in the book presented any convincing evidence of the effectiveness of megavitamin treatment for schizophrenia. Even in the case histories with successful outcomes, megavitamin therapy was usually combined with the use of conventional drugs and psychotherapy. Some of the authors argued that telling the patients their problem is due to a vitamin deficiency, calling it metabolic dysperception rather than schizophrenia, was good psychologically. The review of published research revealed the type of inconclusive picture that characterizes most research in the area.

The findings on orthomolecular treatment were interesting enough to lead to several well-controlled studies by psychiatric specialists who were not part of the health food and megavitamin movement. Several of these studies, which were carefully designed and controlled, showed that treatment with niacin (the most promising orthomolecular treatment) made no difference. In some studies, patients receiving vitamins actually had significantly worse results than those without them. Thus, when the American Psychiatric Association issued a report on orthomolecular psychiatry, it was able to marshal a very impressive body of scientific evidence that did not support the claims of Pauling and his supporters.

Research on mental illness is difficult in many ways. It is hard to be certain of diagnoses and difficult to control for extraneous variables that may influence the results. Pauling could easily have responded to the psychiatrists' report with a moderate statement calling attention to possible weaknesses in the studies and calling for further research. Instead,

Pauling replied with a biting critique, as if he were confronting political opponents instead of scientific colleagues. Rather than conceding that several careful scientific studies had not confirmed his hypotheses, he asked that his critics prove him wrong. As long as the studies didn't conclusively prove him wrong, he thought that he should be given the benefit of the doubt.

Pauling's response left him vulnerable to counterattack, and the editors of the *American Journal of Psychiatry* responded coolly by inviting several prominent specialists in research methodology and drug research to comment on the exchange. These specialists prepared reports that were full of carefully worded, judiciously understated arguments that took Pauling's statements apart, exposed the flaws and gaps in each of them, and left the American Psychiatric Association (APA) looking like a paragon of scientific virtue that had reluctantly found it necessary to correct an errant dotard. Pauling's challenge that his critics prove him wrong demanded a statistical impossibility. There is no way to prove any negative hypothesis conclusively with statistical methods. Pauling had demanded the impossible of his critics and continued to make medical recommendations that were not justified by the data.

After this withering encounter with the psychiatrists, Pauling took a more cautious approach in his next claim for ascorbic acid, his advocacy for its use as a treatment for cancer. He recognized that caution was essential in this case because of the progressive and often fatal nature of the disease, and the fact that effective medical and surgical treatments are available for many types of cancer.

Pauling was excited by research published by the Scottish physician Ewan Cameron and his associates. Cameron administered very large doses of ascorbic acid to patients who had been diagnosed as having incurable terminal cancer. He compared the survival times of these patients with those of similar patients selected independently from hospital records and found the average survival time of those treated to be significantly longer.

These findings could not be presented as a cure for cancer, for the simple reason that almost all the patients died of the illness. From the point of view of research methodology, the study was vulnerable because it is impossible to be certain that the matched cases were strictly comparable. The patients were not chosen at random for experimental and control groups but chosen after their deaths from records of patients in the same hospital over the preceding ten years. Also, it is possible that a "placebo effect" was involved. Survival time may have been extended simply because patients' resistance to the disease was encouraged by being treated by a physician who believed that they had a chance for recovery. It is also likely that some of the patients in the matched control group received an inadequate diet lacking in even the minimum recommended allowance of vitamin C, especially since many were in a weakened state and unable to eat properly.

Pauling's initial advocacy of vitamin C was not based on research he himself had conducted. Beginning in 1973, however, he began to appear as junior author on some of Cameron's publications on the subject. He did not, however, play a role in actually conducting research for these papers. Rather, his role was to help in rewriting them to make them more publishable and then to use the prestige of his name in finding publishers. Increasingly, his publications appeared in the *Proceedings of the National Academy of Sciences.* The advantage of this journal was that Pauling, as a member of the Academy of Sciences, had the traditional right to publish papers without peer review. At times, the editors questioned whether this right should be extended to controversial papers on medical topics, but usually Pauling was successful in having his work published. He and the other contributors had to pay page charges to defray the expenses of publication. Publications in the proceedings are technically labeled "advertisements" to reflect this fact.

Cameron's findings were certainly suggestive enough to justify further research, and Pauling and Cameron were eager to encourage studies in this area. In the March 1979 issue of the influential journal *Cancer Research,* they published an extensive, well-balanced review of the literature on ascorbic acid and cancer. Their article was largely a review of theory and of laboratory findings on the mechanisms by which the body uses ascorbic acid. They had little to report in the way of either experimental studies in animals or clinical trials in humans, and the evidence they reviewed was, in part, contradictory. In studies with guinea pigs, for example, one researcher found that those deprived of ascorbic acid were better able to resist cancer, while another found the opposite. Indeed, one author theorized ascorbic acid was necessary for tumor growth.

The response of cancer researchers to Cameron and Pauling's arguments was lukewarm at first. To many, this approach seemed less promising than other avenues of research. Pauling was angered by this response and was particularly critical of government agencies for not channeling more money into research on ascorbic acid and cancer. He repeatedly suggested that megadoses of vitamin C in healthy persons might prevent cancer—an argument that was not at the time supported by statistical evidence but he considered justified on theoretical grounds, since little harm would come from following his advice even if it weren't successful.

There is a risk in advocating unconventional treatments of this kind. Medical and surgical treatments for cancer and other serious illnesses are often expensive and have unpleasant side effects. There is, therefore, a risk that people who are suspicious of the medical establishment will deny themselves effective treatment for serious illnesses. They may be encouraged by health store owners or "alternative" health practitioners to rely on unproven nutritional cures instead of seeking established treatments that sometimes have well-demonstrated effectiveness.

Linus Pauling was careful never to give such irresponsible advice. In his speeches and writings, he always recommended that vitamins be used together with "appropriate conventional therapy." He also avoided giving medical advice to people on an individual basis, since he did not want to be accused of practicing medicine without a license. However, Pauling did come to the aid of "alternative" health care practitioners who were accused of improperly using megavitamin therapy in place of conventional therapies with proven effectiveness.

In 1984, the California Board of Medical Quality Assurance held lengthy disciplinary hearings in the case of a physician, Michael Gerber, who was found to have improperly used natural remedies in place of conventional therapies. In one case, a woman died of a treatable cancer after being treated with herbs, enzymes, coffee enemas, and chelation therapy. She had refused conventional treatment, as was her legal right, but the board held that it was the responsibility of physicians to "strongly discourage any patient attempts to seek out unproved modalities and nostrums for cancer treatment."[7] In another case, the physician had treated three-year-old twin boys' ear infections with vitamin A and coffee enemas. In both of these cases, the primary allegation was not that the alternative therapies had done harm, but that offering these therapies to patients who were not receiving proven conventional treatments was irresponsible.

In his testimony before the board, Pauling testified that vitamin C "should be given no matter what the decision of the patient is about the conventional therapy that the patient receives." He stated, "I think vitamin C is involved in a very significant way with essentially all diseases." In addition to megavitamin therapy, he also defended the use of coffee enemas, suggesting that the caffeine might be helpful as a stimulant. He said that in his own case, the use of vitamin C had slowed the aging process: "If you get out photographs of me taken fifteen years ago and look at them, when I look at them, I say, I look older than I look now."[8]

The California authorities were not persuaded by Pauling's arguments. They were deferential to him as an expert on chemical matters but did not accept his medical judgments. The result was the same when Pauling testified, in 1992, on behalf of a physician calling himself a "clinical ecologist" in hearings held by New York state authorities. The New York authorities recommended lifting the doctor's license for "gross negligence" and "fraudulent practice."

In 1979, Pauling and Ewan Cameron coauthored a popular book, *Cancer and Vitamin C,* which reviewed conventional cancer treatments and stated that in many cases surgery or chemotherapy is the treatment of choice. It also recommended that cancer patients take large doses of vitamin C. The core of the book was taken directly from Cameron's Scottish research that had been published in professional journals five years before, supplemented with some illustrative case histories from the United States and Canada.

The Cameron data did show some improvement in survival time for terminally ill patients, but Cameron and Pauling had done no additional research to determine whether the findings could be replicated in a more rigorous experiment. Pauling repeatedly urged the National Institutes of Health to fund and conduct such a study, confident that it would definitively validate Cameron's Scottish findings.

Finally, the government did fund such a study under the direction of an eminent cancer researcher, Dr. Charles Moertel of the Mayo Clinic. Moertel designed a rigorous, controlled study in which an experimental group of cancer patients were given large doses of vitamin C, while a randomly matched control group was given a placebo. As is usual in this kind of research, the study was *double-blind,* which means that neither the experimenters nor the patients knew who was getting the vitamin C and who was getting the placebo until the research was over. This procedure was used to control for the *placebo effect,* the possibility that patients will do better as a result of the psychological lift of receiving a new treatment from a physician who expects him or her to benefit from it.

Pauling and Cameron were upset when they learned that the Mayo Clinic study found no significant difference between the experimental and control groups. The paper appeared in the *New England Journal of Medicine* on January 17, 1985. Pauling had asked both Dr. Moertel and the editors of the journal to give him an advance copy so he could prepare his response, but they failed to do so. Pauling argued that the Mayo Clinic trial was not a valid replication of Cameron's Scottish study because the American cancer patients had received chemotherapy prior to taking the vitamin C. Chemotherapy is not so widely used in the United Kingdom, and Pauling argued that the chemotherapy had so weakened the patients that they were unable to benefit from the vitamin C. This is a reasonable criticism: one of the inherent weaknesses of controlled experiments is that the results cannot necessarily be generalized to populations or conditions other than the ones studied.

This was the beginning of a long and bitter conflict among Pauling, Moertel, and the editors of the *New England Journal.* Moertel and his colleagues simply did not believe that the chemotherapy was an important factor. They stated, "Our patients were entered into the study only when they were well past any acute immuno-suppressive effects of previous therapy."[9] In their view, there was no evidence for Pauling's assumption that their immune systems had been permanently weakened by the chemotherapy. They also thought that, if the patients' immune systems had been compromised, then vitamin C should have been effective by helping to restore them.

The journal editors, for their part, expressed their strong preference for double-blind experimental designs, such as that used by Moertel and his colleagues. They gave Cameron's work, which relied on matched controls selected after the fact, little credence. They thought that if Pauling and Cameron had a valid point to make, they should demonstrate it with a controlled study. None of the mainstream American journals was willing to publish Pauling and Cameron's studies, which relied on statistical matching in place of experimental controls.

Pauling, for his part, argued that it would be unethical for anyone who believed in vitamin C to conduct such a study since it would mean denying a medication they believed to be effective. Only a skeptic, who did not believe he knew whether or not vitamin C was effective, could ethically carry out an experiment. Pauling's ethical scruples thus led him to urge that his opponents in this debate do research that he felt he could not ethically do himself. He further insisted that they conduct it according to his precise specifications, which might not coincide with their own priorities.

After a great deal of pressure from Pauling, the National Cancer Institute funded a second study by the Mayo Clinic, this time with patients who had not had chemotherapy. However, the research did not follow Pauling's specifications in every detail. When the results were once again negative, Pauling and Cameron objected to the fact that the patients were taken off vitamin C after an average of only 2.5 months. They thought that the positive effects of vitamin C were lost when the patients were taken off the vitamin. Indeed, they feared that patients might have suffered a rebound effect by being deprived of it after taking large doses.

Pauling and Cameron assumed that the patients were doing well until the vitamin C was withdrawn. However, Moertel and his associates reported that this was not the case. They said that their patients were kept on vitamin C until it was clear that it was not effective or they were too sick to ingest it.

Pauling was extremely angry and upset by the Mayo Clinic results. As Ewan Cameron observed, "I have never seen him so upset . . . he regards this whole affair as a personal attack on his integrity."[10] He thought that Moertel's paper was a case of "fraud and deliberate misrepresentation" and consulted his attorneys about the possibility of filing a lawsuit, either a libel suit or a class action suit on behalf of all cancer patients in the United States. His attorneys advised against filing any kind of suit, arguing that the courts would be reluctant to intervene in a scientific argument, and that if they did get involved in a case involving medical treatment issues they would tend to side with the medical doctors.

The *New England Journal of Medicine* offered to publish a five-hundred-word letter to the editor from Pauling, but he insisted on sending two longer letters. He also insisted that they explain to him how it came about that they had published a fraudulent paper. When the editor, Arnold Relman, did not answer this inquiry, Pauling sent the following letter:

> Some time ago I wrote to you, asking for information about the process by which this fraudulent paper came to be accepted for publication in your journal. You have not answered my letter. I hope that you will do me the courtesy of answering my letter. Your continued failure to do so would indicate that you also are involved in this conspiracy to suppress the truth.[11]

Ewan Cameron thought that Pauling was overreacting. His assumption was that they were dealing with fools, not knaves. However, he also became frustrated by the repeated failure of the *New England Journal* and other mainstream medical journals to publish his nonexperimental research. His observations in a letter written in 1987 summarize Pauling's core beliefs about the medical profession:

> In my long, pleasant and productive association with Linus Pauling, we have had one matter of constant disagreement. Dr. Pauling, like many of the American public, sincerely believes that the medical profession is waging some kind of monstrous conspiracy to maintain its prestige, its power, and its free-earning capacity at the expense of the public. I have argued vehemently with him that this is not so, and that our profession may well be too conservative and slow to accept new ideas, but is always motivated by the highest ethics and ideals. This latest episode makes me wonder whether it is I who have been overly naive these many years.[12]

Pauling's angry response was not limited to impulsive, off-the-cuff remarks or correspondence but was also expressed in his book *How to Live Longer and Feel Better,* which relies on the same techniques as his earlier books on vitamin C: selective quotations from the literature bolstered with anecdotal evidence and disparagement of studies that disagree with his conclusions. In the book, he describes the Mayo Clinic studies as "outrageous" and "fraudulent" and observes:

> The Mayo Clinic doctors have refused to discuss this matter with me. I conclude that they are not scientists, devoted to the search for truth. I surmise that they are so ashamed of themselves that they would prefer that the matter be forgotten. The Mayo Clinic used to have a great reputation. This episode indicates to me that it is no longer deserved.[13]

At the Linus Pauling Institute, much of the hands-on research activity was conducted or closely supervised by Arthur Robinson. Robinson was an experimentalist whose talents complemented Pauling's strengths as a theorist. He had begun his association with Pauling when he was an undergraduate laboratory assistant, and he had managed his own and Pauling's laboratories at the University of California in San Diego and at Stanford University. He was the kind of scientist who spent long hours in the laboratory, making sure that experiments were conducted properly.

Robinson was about the same age as Pauling's sons, and many people believed that, on an emotional level, he was a surrogate son for Pauling. As a distinguished professor, Pauling had naturally worked with many graduate students who did routine work in exchange for a chance to be mentored by him. This can be a tricky emotional relationship, especially if it is prolonged. There is an inherent conflict between the student's need to become more independent and the mentor's need for a protégé to follow in his footsteps. Of all Pauling's students, Arthur Robinson was the one with whom he had the closest and longest relationship. When asked at a legal deposition "Have you ever viewed Arthur Robinson as being like a son to you?" Pauling answered:

> Well, that would make me into a Turkish pasha. I have so many students that I have similar feelings about. I've thought about many in the way that I do about a large number of young people who have been associated with me.[14]

As Pauling conceded at the deposition, however, there were no other students with whom he had started an institute. Indeed, Pauling and

Robinson's association at the Linus Pauling Institute was much like a "father and son" business, a much closer professional working relationship than Pauling had with any of his own sons. Both invested their personal funds and energies in starting it and helping it through fiscal crises.

For Pauling, who had been placed on emeritus status with no pay by Stanford, the institute promised a continuing income and a chance to build an institution that would carry on his work after he was gone. Robinson gave up the opportunity for a tenured university position in the hope of building a base where he could serve humanity by doing health research independent of government and institutional constraints. As Robinson saw it, "Because of [Pauling's] age, I felt I should do most of the work building the place up because I'd have longer to enjoy it."[15]

One of Pauling's favorite observations was that he believed in "leaving the hard jobs to others," and the building up of the institute was no exception. Robinson worked very hard, putting in sixty- to eighty-hour weeks. He developed instruments and computer programs needed to measure the content of various chemicals in body fluids such as blood and urine. Primarily through Robinson's efforts, the institute received federal research grants totaling over six million dollars to support this work. The overhead on these grants provided much of the basic funding for the institute in its initial years. Pauling devoted himself primarily to writing popular articles, traveling, giving speeches, and writing articles on chemical topics unrelated to medicine at his Big Sur ranch.

Initially, Pauling was president, director, and chairman of the board; Robinson was vice-president, assistant director, and treasurer. Two years later Robinson became president and director. Each was made a research professor with tenure. Pauling had tenure for life; Robinson until age sixty-five.

During his bachelor years, Art Robinson spent Christmas and other holidays with Linus and Ava Helen and was often a guest at their Big Sur ranch, to which only family and intimate friends were invited—not that everyone valued such an invitation. Rae Goodell observed that you had to be prepared to read encyclopedias and to talk only about serious matters, and at all costs avoid idle "chitchat," in Pauling's presence. Collecting encyclopedias was Pauling's hobby. Robinson came to know the Paulings' sons and daughter and was seen by others as almost a member of the family, especially after his own parents, Zelma and Ted Robinson, died tragically in 1966–67. He was their only child.

During these years, Robinson received several hour-long telephone calls from Peter Pauling in London in which Peter expressed his anger at his father. He stated that he had problems visiting his parents in California, because his father always had to have his way about everything. He expressed similar sentiments to the authors of this biography, cautioning us, "My father is not the great man you think he is."

When Arthur Robinson and Linus Pauling were founding the institute, Robinson reports, "Peter came to see me and said my father is just using you and my mother and I are both worried about you because he is just using you." At the time when Pauling's professorship at Stanford was being converted to emeritus status, one of the alternatives that Robinson considered was taking his grants back to San Diego and pursuing his collaborative research with Pauling from a base there. Later, Robinson very much regretted not taking the advice of friends who urged him to do this. Hicks remembers hearing Robinson refer to Pauling sardonically as the "Great Man" or the "Pied Piper."[16] At the time, however, the appeal of working as an equal partner with Linus Pauling was too powerful to resist.

In 1972, Robinson married and began spending less time with the Pauling family, but their relationship continued to be close. Laurelee and Arthur Robinson soon began what was to be a large family, whom they intended to educate at home, where they could maintain an environment supportive of their fundamentalist Christian beliefs and high intellectual standards.

From the beginning Laurelee also contributed to the institute, working as a volunteer during the early days when her husband and Pauling worked without salary and had to pay staff out of their own pockets. She continued to do institute work at home after her children were born, because she did not like to leave the children with a sitter. She performed almost all of the computer programming for the research done at the institute, although her master's degree was in biochemistry.

For Ava Helen and Linus Pauling, his work at the institute was a pleasant kind of semiretirement since they were free to spend about a third of their time at the Deer Creek Ranch, another third traveling in the United States and abroad, and the rest working at least part of each day at the institute. Also, they were well paid after the institute became well established with Linus's approximately $80,000 salary including benefits augmenting their other income from royalties, honoraria, and investments. Robinson's salary (as president, director, treasurer, and tenured research professor) eventually rose to $48,400, including retirement benefits. As the years passed, Pauling took little interest in activities other than his own: his popular articles on vitamin C, his lectures, his ongoing research on chemical bonding. For recreation, Pauling continued his lifelong diversion of searching through journals in crystallography looking for errors and writing corrections, as happy as a child reading comic books.

Robinson had been willing to spend half of his time as an administrator during the early years, to help establish an institution that would afford him the leisure to devote himself entirely to research in his own middle age and older years. The only fly in the ointment was the institute's persistent financial problems. Robinson spent a lot of time cultivating wealthy individuals, who wanted personal contact with the institute before giving large sums of money. He began to feel that he would never get his own research done if he had to travel around the country with a briefcase smiling at wealthy people. In the winter of 1973, the institute reached a low point financially. If it had closed at that time, Pauling and Robinson would each have had to chip in $80,000 of their own money to pay off debts. Fortunately, a California businessman, Stephen Kahn, came through with a $400,000 gift.

Pauling was inclined to leave the fund-raising to others, but he enjoyed being able to offer jobs to people, especially young people with progressive views who were having trouble adjusting to the establishment. One young man, who ate so much garlic that no one could work closely with him, set up residence in a purple van in the parking lot. He had little academic training and had been unable to adjust to graduate school, but Pauling asked Robinson to find him something to do. He ended up being listed as coauthor on several papers with Pauling, then grumbled that Pauling had taken credit for his work.

In Robinson's view, Pauling had a weakness for favoring the underdog. If he had three or four postdocs working for him, Robinson thought he would give more time and attention to the weakest one. He often favored giving promotions to the people Robinson thought were the least productive. Pauling's instincts were those of a compassionate liberal. He had a soft spot for people who were in financial need or were having personal or career difficulties.

Robinson, by contrast, was a free market conservative and seemed less sympathetic to the underdog. He thought he could relieve himself of the burden of fund-raising duties by hiring a professional to do it. He hired G. Richard Hicks, a former stockbroker who was a health food enthusiast. Hicks told Robinson that he had made enough money and now wanted to do something for humanity. Hicks was euphoric about Pauling's theories about vitamin C and Pauling seemed to like him.

Unfortunately, Hicks's efforts were not sufficient to keep the institute going, so Robinson continued to bear much of the burden of fund-raising. In particular, he planned a massive junk mail campaign, based on lists purchased from *Prevention* magazine. This was expensive but successful. The institute was inundated with mail from donors, some of which had to be answered personally. Hicks was not familiar with data processing, so Robinson took charge of setting up systems to manage the bookkeeping and correspondence.

At the same time, Robinson was pursuing his dream of relocating the institute to a rural area. He found a plot of land on a plateau in Oregon, midway between the University of Oregon and Oregon State University. Some of the staff members grumbled about having to leave the San Francisco Bay area, but Pauling seemed interested. Perhaps it appealed to him to build a tangible monument to his ideas in his home state.

Robinson felt that the five years of hard effort he put into the institute were beginning to pay off. They had built a significant analytical laboratory and the research was going very well. A list of sixty thousand donors seemed a secure guarantee for the future. Although there had been no breakthroughs against cancer or other dread diseases, the institute's reputation was bolstered by over eighty publications, most of them authored or coauthored by the two founders. He believed that several of their most ambitious experimental programs were about to bear fruit. Bound copies of the research were provided to the donors and associates.

Optimistically, Pauling and Robinson and the other trustees, now six in number, offered appointments to several new faculty members. At Robinson's suggestion, 50 percent of the research funds were allocated to Pauling directly for the remainder of his career and the other 50 percent were shared by the other faculty including Robinson. Now that the institute was expanding, Pauling was eager to bring Ewan Cameron to Palo Alto. He went to Scotland to invite him personally. Robinson states that he was also eager to recruit Cameron and wrote a warm letter reinforcing Pauling's invitation. Art and Laurelee had entertained the Camerons on their previous visit and had been entertained in the Cameron home in Scotland. Pauling, however, became convinced that Robinson secretly opposed appointing Cameron and was working behind his back to sabotage the appointment.

Cameron did seem to have some misgivings about joining the institute. He was having some success in setting up research projects in hospitals in Great Britain, and he was not sure Pauling would be able to establish a similar facility at the institute. Dr. Cameron, a canny, reticent Scot who towered above Pauling, was not given to grandiosity and was suspicious of excessive spending. He told Robinson he was skeptical when Pauling offered him fifty thousand to seventy-five thousand dollars a year for as much as ten years, seemingly on the spur of the moment. When he finally consented to join the institute, it was for only one year.

Pauling seemed pleased with Robinson's continued experimental work on health profiling, collecting detailed data on chemicals in volunteers' bodily fluids, and with their later work on the incidence and severity of skin cancer in hairless mice exposed to massive amounts of artificial sunlight. Robinson says that he designed and directed this project, with only minor input from Pauling. Certainly he was the one who watched over it on a day to day basis. Pauling, however, later claimed that Robinson was functioning as his assistant in this project. It was funded jointly out of Pauling and Robinson's research accounts at the institute.

They tested out a number of ideas from the health literature to find out whether they would help to build resistance to cancer. Some of this research, to which Pauling had given scant attention, found that mice have less squamous cell carcinoma when given doses of vitamin C equivalent to about 50 to 100 grams daily for an adult human.

Robinson had become impressed by the work of Eydie Mae Hunsberger, who claimed in a popular book that she had cured her cancer with fresh fruits and vegetables. He tried giving a group of cancerous mice no treatment except a diet of raw fruits and vegetables, and they improved as much as did animals on high doses of vitamin C, a result Pauling called "a bombshell." When Robinson combined the raw fruit and vegetable diet with high doses of vitamin C, the results were extraordinary; the cancer lesions were suppressed over 3,000 percent, or about thirty-five times more than were those of the controls. This meant that the mice lived longer before suffering from cancer. Eventually, however, all of the mice were afflicted.

Robinson and his coworkers were excited about their findings and explored the effects of systematically removing one or another fruit or vegetable from the diet. The positive results were reliable and reproducible. Adding protein in the form of nuts and seeds to the raw fruit and vegetable diet had a negative effect. This suggested that protein might sometimes promote cancer, at least in mice, and perhaps in humans, so Robinson immediately designed a series of experiments to follow up this possibility.

The seeds of a major crisis were planted when Robinson reported findings from the hairless mice experiments that conflicted with Pauling's ideas about vitamin C. In one experiment, Robinson found that mice given the human equivalent of 10 grams of vitamin C per day while eating a conventional diet actually had *more* lesions than the untreated controls. This is the dosage that Pauling recommended for most adults, and that he and Ava Helen routinely took.

Although Pauling had never suggested that anyone should use vitamin C as an alternative to conventional cancer treatments, he had always minimized another criticism of his recommendations, the possibility that large doses of vitamin C might be an irritant and might actually increase the risk of cancer. This possibility may have been particularly troubling for Pauling at this point because Ava Helen had been diagnosed with stomach cancer in 1975 after years of taking large doses of vitamin C. Neither conventional nor alternative therapy was able to cure her disease, and she died five years and three months later. Of course, Pauling could take comfort in the possibility that vitamin C prolonged her life span after diagnosis, but there is also the nagging possibility that the massive doses had some role in causing the cancer.

Robinson did not think of his work as fundamentally contradicting Pauling's theories, but simply as elaborating on them and suggesting additional avenues of research. Pauling, however, came to view him as a threat. One day he called Robinson into his office, trembling with anger, and said, "If you have something to say about a person, you say it to his face." He then accused Robinson of having told Ewan Cameron that he was senile. Robinson denied the charge vehemently and asked that they call Cameron immediately to set the record straight. Pauling asked for a delay, then asked Robinson not to call Cameron for fear that such a call might strengthen his resistance to joining the institute.

Robinson was convinced that Richard Hicks had told this story to Pauling, a fact Pauling later confirmed at a legal deposition that was part of Robinson's lawsuit.[17] At his own deposition, Hicks stated that Cameron had told him that Cameron "felt that Dr. Robinson thought Dr. Pauling was senile" and that "he did not believe that Dr. Robinson believed in the work that Dr. Cameron and Dr. Pauling were doing." According to Hicks, Cameron also thought that Robinson didn't want him to join the institute and was, in fact, offering him financial inducements to remain in Scotland.

Robinson vehemently denied ever having said or thought that Pauling was senile. Indeed, he said that he repeatedly defended Pauling against this accusation when it was made by scientists critical of his views on vitamin C. He suspected that Hicks invented the entire incident in an attempt to sabotage him. It is also possible, of course, that Cameron misunderstood or misinterpreted something he said.

Robinson is eager to talk at length with anyone who is interested in the problems at the institute, or any other aspect of his relationship with Pauling. His earnest demeanor and complete apparent openness, combined with his strong religious beliefs, make him a believable informant. Journalists who wrote about the incident generally seemed to believe his account.

Pauling and others still affiliated with the institute, by contrast, generally refused to discuss the whole affair. Their testimony at the legal depositions, however, provides a good account of their views. The account of the incident that follows is based on three well-researched magazine accounts,[18] a lengthy interview with Robinson, and a careful reading of hundreds of pages of legal depositions.

Robinson was furious with Hicks. He had already decided that Hicks was incompetent and irresponsible as a fund-raiser. Now he was convinced that Hicks was conspiring behind his back. When Pauling was unavailable to meet with him and Hicks to discuss their dispute, Robinson took it upon himself to fire Hicks. This was within his formal legal authority as president of the institute, but went against the organizational reality that the ultimate power lay in a board of directors that was loyal to Pauling. The board members thought that Robinson was acting irresponsibly in taking this action without Pauling's approval and in rejecting Pauling's suggestion that operational authority be placed in a governing committee.

Robinson, however, stubbornly stuck to his guns. He was president of the institute, and he had been an equal partner with Pauling in founding it. He had worked night and day for five years to build it up, and he felt he was responsible for most of the research going on within its walls. Hicks had been hired to assist him, Hicks reported to him, and Robinson had the clear authority to fire him. He called Hicks into his office, telling him that he would be willing to take the time to work out their difficulties if Hicks would be open with him and make an honest effort to do so. When he became convinced that Hicks was not willing to do this, he fired him.

Hicks didn't stay fired for very long. The board, with Pauling's approval, quickly reinstated him and resolved to get rid of Robinson instead. The most dramatic confrontation for Robinson occurred when Pauling called him into his office. According to Robinson, Pauling said, "I demand that you leave the office, turn over all your research to me, and never come here again." Robinson asked him, "Why do you say that?" Pauling replied, "I've found out. If you do not do what I ask I'll have to tell people what I know." Robinson asked, "What is it?" Pauling wouldn't tell, claiming, "If I discuss it I will feel ethically bound to tell others." When Robinson said he wouldn't leave without some reason being given, Pauling pointed to a copy of *New West* magazine and threatened that if Robinson didn't leave, "Magazines will write bad articles about you; you will never work in science again; I will see to it. I will see that you never work in science again and I'll tell what I know."

Robinson's first thought was that someone must have told Pauling a monstrous lie about him. For a year and a half, he kept believing that if only he could find out what the accusation was, he could prove that it wasn't true and get back into Pauling's good graces. He never did find out what Pauling was threatening to expose. At a board meeting later, Pauling entered holding a document that he described as a "bill of particulars" against Robinson. When Robinson insisted that his attorney be present, however, they suspended the meeting. When it reconvened, they declared that the first part never happened and rewound the tape that had recorded the proceedings. After consulting with their lawyer, the board simply fired him without stating their reasons. Robinson later felt that he made a mistake in not getting the bill of particulars on the record at that time. He wanted desperately to set things right with Pauling, but to no avail.

This began a lengthy, expensive, and stressful legal confrontation between Robinson and the institute. The board could remove him as president at its pleasure, but they conceded that he had tenure as a research professor. There was no document defining precisely what tenure at the institute meant, but the assumption was that it was similar to tenure at a university. The essence of academic tenure is to protect faculty members from being fired because their research findings upset some established doctrine. Under such a policy, once Robinson was relieved of his administrative duties, he should have been able to pursue his own research. Robinson, however, told us, "from the the first moments of his attack, Linus's demand was that I resign all positions at the institute, turn over all of my research to him, and physically leave the premises. He never deviated in this demand and was never willing, even when asked by the trustees, to turn over any of my research data to me."

When it became apparent that the board could not easily fire Robinson, Pauling suggested that he should take a year's leave of absence with pay as a sort of sabbatical. In fact, the institute did continue his pay for about a year, when their lawyers told them that they couldn't simply stop paying him. But Robinson was denied access to the laboratories where his experiments had been conducted. He wasn't allowed inside the building. Nor could he get access to about five hundred computer tapes on which he had stored his research data. The institute never actually said he couldn't have the tapes, but they somehow never could find them. They also destroyed a cabinet full of biological materials that Robinson had collected over the years, on the grounds that some of them were radioactive. Robinson claims that the radioactivity was a minor residue of standard tests and did not pose a problem. The mice were killed, presumably after an independent pathologist examined them.

The institute's lawyers claim that Robinson was offered the opportunity to take his samples home. At one time they said he could enter their offices for two hours and photocopy his computer printouts, which had been stored in numerous file cabinets. At twenty-five cents a copy, this would have cost Robinson a fortune and would not have been a practical solution to his need to complete his research.

The institute's major fear seemed to be that Robinson would somehow sabotage the computer system, or use the mailing lists to send hostile mailings to the contributors. Laurelee had done all the computer programming, and they weren't even sure they could get it to work without her. They hired an independent computer consultant to work on it, but he was afraid he might lose the files and turned to Laurelee for help. Laurelee went in with him and backed up the mainframe files so they would have a backup, unbeknown to the institute staff. Unfortunately, she was unable to copy the five hundred smaller tapes containing their research findings onto a large tape and take them home.

Robinson did get four file drawers of his correspondence but was talked into returning them to the office. Emile Zuckercandl, a biologist who was on the board and worked closely with Pauling, pleaded with Robinson to return them before Pauling found out they were gone. Everyone at the institute was fearful of Pauling's wrath. Robinson says that Zuckercandl promised "on my sacred honor" that Robinson would have access to the files, but he never saw them again.

In the legal struggle, Arthur and Laurelee were up against one of San Francisco's largest legal firms. The institute spent about a million dollars in legal fees, and they expected to be able to wait the Robinsons out. The Robinsons first tried conducting their defense themselves but weren't able to do it and had to hire a lawyer. They moved to rural Cave Junction, in the far south of Oregon, and set up a small laboratory of their own. They purchased a computer, assuming that they would get access to the data from Robinson's five years of experiments.

When the data never materialized, they turned their computer skills to analyzing data from the commodity markets in Chicago. At this time, computer analysis of financial markets was in its infancy, and few people were doing it. They succeeded in making about $500,000 over the course of a year, most of which was poured into the legal confrontation.

Finally, the lengthy depositions were coming to an end, and it appeared that the suit would actually be going to court. By this time, Robinson was pretty much convinced that his research materials were lost for good. He agreed to a settlement of about $500,000, which was the amount which they estimated would settle their legal and other debts and allow them to start fresh in Cave Junction. That part of the settlement was for libel and slander, Robinson felt, amounted to a moral victory.

The Robinsons set up the Oregon Institute of Science and Medicine in Cave Junction and set about raising their six children. They kept enough farm animals to qualify for reduced taxes as a farm, and Laurelee devoted herself to home schooling the children. With time, the farm became a debt-free, self-sufficient unit and they were able to carry on some research on nutrition and cancer. They also pursued their political commitments, forming an organization called Fighting Chance that advocated a stronger civil defense system for the United States. They wrote the 1988 Republican party platform plank on civil defense.

In November 1988, Laurelee died after an illness of less than twenty-four hours. She was diagnosed with idiopathic acute hemorrhagic pancreatitis, an extremely rare condition for a woman her age. Arthur was devastated, unable to understand why God had allowed Laurelee to die in the prime of her life with all her dreams before her. He continued to live in Cave Junction with the children, developing a home study curriculum in which the children worked through mathematics lessons on their own, then were free to read from great books. The children flourished academically on this regimen.

Pauling, for his part, was reticent about discussing the episode with

Robinson. When pressured by an interviewer to respond to Robinson's allegations, Pauling responded:

> I don't want to attack Dr. Robinson. I don't know if Dr. Robinson is attacking me. Dr. Robinson has filed suits against the Institute. I don't know the lawyers, the statements that are made in these suits. I recognize this legal terminology [may not be] a personal attack against me. I can't claim that he was the best of my students in academic ability, but he was a remarkable man.[19]

Pauling, of course, had participated in many lawsuits and he understood that statements may be made for legal purposes that do not represent people's actual feelings. His tone of voice conveyed a sort of paternal concern for Robinson and hope that he would mend his ways, learn from his experiences, and go on to have a successful career.

At the same time, he consistently referred to Robinson as a junior partner who assisted him in his work, not as a colleague and partner. In discussing the mouse experiments with a reporter from *New West* magazine, Pauling said:

> Dr. Robinson has been discussing this work as if it were his research, but there is good documentary evidence that it is my work and I asked him to help with it.
>
> He doesn't have any special reputation as a scientist to uphold. He can go around making statements without damaging his reputation. There are thousands of scientists with his ability and achievements.

This was an unkind thing to say about a man who had been his closest collaborator for the preceding ten years. The reporter from *New West,* Jerry Carroll, sought out a response from "a famous biochemist friendly with them both," who commented: "That's another typical Pauling statement. Art's an exceptionally capable and outstanding scientist."[20]

Robinson objected vehemently to Pauling's claim that he was principally responsible for the research at the institute. Pauling was able to make the claim that he was bringing in research money because he was listed as principal investigator on a number of large grants. However, Robinson claimed:

None of these federal research proposals, none of the research reports, and none of the resulting fifteen published papers were written by Linus. He did not originate, direct, or carry out any experimental or theoretical research relevant to these federal research projects. I listed his name on some proposals and publications for administrative and political reasons, but I was never successful in obtaining Linus's actual participation in the work. I was principal author of twelve of the research publications, my co-workers of three, and Linus of none.[21]

During the long dispute, Robinson kept trying to get Pauling to make a straightforward statement of the reasons he had been fired. Most observers, including the two reporters who investigated the incident most fully, concluded that the primary reason was the disagreement about the experiment with the mice. In a response to an article we published in *Antioch Review* in 1980, Pauling stated, "It was not because of any dispute about the mice but rather for other reasons that the Board took these actions."[22]

Unfortunately, Pauling would never state what these "other reasons" were, as if there were some dark secret that could never be revealed. Finally, on September 18, 1980, Pauling responded to a subpoena to testify at a deposition in Robinson's lawsuit. Here, in response to a full day of probing questions from Robinson's lawyer, Pauling gave his account of the conflict with Robinson under oath.

First of all, Pauling explained why he was reluctant to turn over scientific information on the mouse experiments to Robinson. He claimed that Robinson was stealing and distorting *his* research:

I would say that I have a scientific reason for being concerned about giving additional information to Dr. Robinson. He used the information to damage the institute by publishing some of these results himself without consultation with me, whose work it was that he was reporting on, my project, and in a quite unscientific way that he did not take into proper consideration the statistical significance of the observations perhaps because of his lack of familiarity with biostatistics.

In view of the situation that exists in which he has assumed an antagonistic position toward me and the institute, I feel that from the scientific point of view he might well make an effort to use additional infor-

> mation, whatever there is, in an unscientific way just to cause further damage to me and to the institute.
>
> So, from a scientific point of view, I would say that I need to know now just what it is that you are asking for that he does not have. And I should like to have some assurance that he would not misuse the information.[23]

During the course of the depositions, Robinson's lawyers tried to collect evidence to prove that Pauling was not closely involved in the research with the mice. It is clear that Pauling did not involve himself in handling the mice on a day-to-day basis, although he may have observed them occasionally. But there was no way that the witnesses could conclusively demonstrate whose research it was, since the two of them discussed it in private. Robinson claims that he received only minor suggestions from Pauling, but there is no way to document what went on in their private conversations. Pauling claimed, for example, that he got the idea of using raw fruits and vegetables from a book by Max Gerson that he had read in the 1950s. Robinson claims he got the idea from the Hunsbergers, health foods enthusiasts without scientific training.

Pauling objected to the fact that Robinson "had made arrangements with the Hunsbergers to collaborate with them without my knowledge. And this collaboration had, I thought, unfortunate consequences."[24] Robinson, for his part, insisted that as a research professor he had the right to collaborate with whomever he wished.

Pauling insisted that it was not his intention to destroy Robinson's career: "Not at all, I've always worked to bolster him up and to help him to become a good scientist."[25] He stated that Robinson was "an able scientist. He has some remarkable qualifications, the ability to get things done. That is one of his remarkable qualifications."[26]

Why, then, was it necessary to fire Robinson? Pauling said that, although he personally had always gotten along well with Robinson, other people found him difficult to get along with. Over time, he felt that Robinson was increasingly making remarks such as "If you say so, Dr. Pauling," which gave Pauling the impression that "he was accepting my suggestions but didn't agree with them."

The worsening relations between them came to a head over the issue of the relationship with the Hunsbergers. The Hunsbergers were running a clinic in Santa Cruz where people could learn about their raw fruit and vegetable diet. Robinson seemed to want to establish some kind of organizational link between this clinic and the Linus Pauling Institute. The Hunsbergers, in addition to being health activists, were quite wealthy and might have helped out financially. Pauling, however, did not want to associate so closely with nonscientists. When Pauling discussed this with Robinson, Pauling said, Robinson

> became very antagonistic to me. And it was clear that he was not making an effort to continue to be friends with me. He resisted my request to him to break off relations with the Santa Cruz institute, whatever they were, with the Hunsbergers at first saying that he's a professor at the institute and he has a right to.[27]

Pauling also believed that Robinson was interfering with his efforts to recruit Ewan Cameron to the institute. Robinson was not following up on some of his research suggestions. Pauling, therefore, was unhappy with Robinson's performance as president of the institute because of policy differences, especially with regard to the Hunsbergers, and because he thought Robinson was not getting along well with him and with others.

Pauling tried to regain control of the institute by setting up a three-man executive committee to make key decisions, consisting of Pauling, Robinson, and Hicks. Robinson countered with a suggestion that there be a larger executive committee, most of its members board members not resident at the institute.

While these negotiations were going on Pauling became angry because he heard that Robinson had told Ewan Cameron that he was senile. When Robinson fired Hicks, without consulting Pauling, Pauling decided it was time to have him removed as president.

This was an action fully within the discretion of the board. The legal conflict came about because Robinson was also fired from his position as a tenured research professor and denied full access to his research materials. With regard to that firing, Pauling testified:

> The only reason that I asked the board to terminate him also as research professor was that I had told him that if he did not act in a sensible man-

> ner by resigning without making trouble, resigning as president and director, I would then ask the board to remove him from all connection with the institute. So I was living up to my word, the statement to him. And I did not advance any charges of misconduct against him at that time."[28]

Pauling has never revealed any of the mysterious charges to which he alluded when he demanded that Robinson leave the institute. Although Pauling had threatened that magazines such as *New West* would write critical articles about Robinson if he did not resign, the outcome was that *New West* published an article titled "The Perils of Pauling," which reflected badly on Pauling, not on Robinson. A lengthy article in *Barrons* was also unflattering to Pauling and the institute. The authors of both of these articles concluded that Robinson was fired principally because of Pauling's desire to suppress the findings from the mouse experiments.[29]

Although Robinson's lawsuit was settled, the controversy over the research with the mice continued. On January 29, 1980, Pauling sent Robinson a letter informing him that he had finished the analysis of the four-year study of cancer in hairless mice and asking Robinson to join the list of junior authors who would follow him in the list of authors. Robinson reacted with indignation, claiming that Pauling had appropriated his research and written it up in a way that was biased and incomplete. In his view, Pauling's version of the paper glossed over the fact that increased rates of cancer resulted from certain dosage levels of vitamin C. Robinson claimed that Pauling was now attempting to take credit for the experiments with raw fruits and vegetables even though Robinson performed that experiment without even telling Pauling about it. He also claimed that Pauling had obscured the unwanted results by using a linear scale instead of the usual logarithmic scale in plotting the dose-response curves; the linear scale, while accurate, made the discrepant data look smaller on the graph. He wrote to Pauling objecting to the publication of the article and pointing to the problems he saw in it.

Robinson later found out that Pauling had submitted the paper to *The Proceedings of the National Academy of Sciences* (*PNAS*). Robinson observed that in the manuscript Pauling

> extrapolated some of the results and concluded "that the raw-foods diet with about 5% added vitamin C would provide essentially complete protection against skin cancer in irradiated hairless mice under the conditions of these experiments." Since he had not been involved in the work, he did not know that raw food with 5% added vitamin C is lethal under those conditions. They would avoid cancer only in that they would not live long enough to acquire it. This diet corresponds to an intake of about 400 grams of vitamin C per day for a human.[30]

Robinson wrote to *PNAS*, and they refused to publish the paper. However, Pauling and four of his colleagues did publish the results of a number of mouse experiments in the *International Journal for Vitamin and Nutrition Research* in 1981.[31] Robinson was not listed as one of the five authors, but an acknowledgment stated, "Dr. Arthur B. Robinson rendered much assistance with these studies."

This paper did not mention the raw fruit and vegetable experiments but reported results for different dosages of vitamin C. The authors note that there was a higher rate of cancer with 0.3 percent vitamin C in the diet than with no vitamin C at all. They "surmised that this observation resulted from statistical fluctuations" and decided to do an additional experiment just to make sure. The follow-up experiment showed no difference between the experimental and control groups.

Pauling and his colleagues conducted a sophisticated statistical analysis on their data set and concluded that on the whole vitamin C tended to decrease the incidence of malignant lesions. One of the graphs includes the parabola to which Robinson objected. The text explains that the parabola "has no significance other than to guide the eye." Robinson says, "Yes, to guide the eye away from the findings."

The second experiment and the overall pattern of variation in the data are consistent with Pauling's suggestion that the initial finding that generated so much controversy was a statistical fluctuation. Furthermore, both Pauling and Robinson agree that experiments on hairless mice cannot be readily generalized to the human population, since the mice create their own vitamin C internally, and it is possible that they simply compensate for increased C in their diet by producing less themselves. However, Robinson and his colleagues Arthur Hunsberger and Fred Westall continue to contest Pauling's conclusions and published data in 1994 showing that dietary variation can affect cancer rates in hairless mice.[32]

Scientifically, the whole controversy over the mice was much ado about very little. The original experiment was perfectly valid and perhaps even potentially important, but Pauling was right in insisting that it needed to be replicated to determine whether or not it was a statistical fluctuation—as indeed it may have been. There were some dramatic differences between groups of mice at certain points of time. If you inspected the mice at just the right week, as these biographers happen to have done on a visit to the institute, some batches would be covered with ugly lesions, while others were perfectly clear. Within a few weeks, however, all the groups succumbed to the cancer. Ironically, perhaps the last word was expressed by one wealthy potential donor, who adamantly refused to contribute funds to the institute as long as they were experimenting on helpless animals. She suggested they should experiment on human prisoners instead.

The intense emotional conflict between Robinson and Pauling contrasted sharply with the triviality of the scientific and practical issues involved. The conflict occurred at a time when the institute was on a sound financial basis for the first time, thanks to the successful direct mail campaign. If both parties had been able to settle their dispute rationally, they could have completed more mouse experiments to resolve the issue scientifically and done a great many other things as well.

When intelligent, capable people behave in irrational, self-defeating ways, those around them offer a range of explanations. Some suggested that Pauling simply did not need Robinson anymore, now that the institute was on a firm footing, and perhaps wanted to bring in his family. But none of Pauling's family wanted to work in the institute; nor did any of them wish to harm Arthur Robinson. Others suggested that Hicks had manipulated and deceived Pauling with a devious plot. But, if so, Pauling must have been willing to be deceived, since he could have easily checked out any accusations Hicks made.

Several people observed the parallels between Pauling's difficulties with his son Peter and his conflict with Robinson and suggested that perhaps Robinson was a proxy for Peter in Pauling's unconscious mind. This kind of speculation is, of course, unverifiable, but the parallels are there, as they are likely to be in any situation in which a young man and an older man join in such an enterprise. Robinson had a strong need to assert his independence, and Pauling insisted on maintaining control. Given their strong differences in political philosophies and working styles, perhaps the greater puzzle is how they stayed together so long.

To complete the psychoanalytic interpretation, of course, one would have to examine Arthur Robinson's biography and motivations, including his difficulties in dealing with his own parents' deaths. With hindsight, Robinson recognized that he made a major life mistake in following the "Pied Piper" rather than pursuing a university career. His apparent belief that he could win control of Linus Pauling Institute away from Linus Pauling suggests that his keen intelligence was clouded by emotional factors. After the breakup with Pauling, he was obsessed with the lawsuit for many years. Even after it was settled, he became involved in a prolonged and expensive conflict with his own attorney, suing him for copies of documents related to the suit.

The Robinson affair was a sad episode in Linus Pauling's life. Some of his other collaborators had believed that they should have received more credit for joint work. But there had been no other case in which a close collaborator's career had been damaged, or a relationship had broken up in open acrimony and lawsuits. Pauling's previous suits had all been with political opponents, not close associates.

9

Alone at Big Sur, 1991–1994

In December 1991, at the age of ninety, Linus Pauling was diagnosed with cancer of the prostate. He did not interpret this diagnosis as contradicting his theories, however:

> I've never contended that Vitamin C in large doses will protect you absolutely against any disease . . . many men die of prostate cancer at age 60 or 70. Taking Vitamin C may mean that my curve shifted 20 years.[1]

Pauling remained convinced that vitamin C was almost totally effective against the common cold. Pauling routinely took about 10 grams of vitamin C a day. When he was travelling and felt tired or thought he felt a cold coming on, he sometimes increased his dosage to as much as 50 grams. The fact that the 10 grams had not been sufficient in the first place did not shake his confidence in the vitamin. And when he had symptoms that other people might think of as a cold, such as a bad runny nose or a cough, he attributed them to allergies.

As his life drew to a close, Pauling had many reasons to feel satisfied. In a reflective interview with the Japanese writer and Buddhist philosopher Daisaku Ikeda, he reviewed his accomplishments and some of the lessons he had learned. Although he still considered his work on the hybridization of orbitals, published in 1931, to be his greatest scientific achievement, he was also deeply satisfied with his campaign against nuclear testing and his contributions in orthomolecular medicine. He believed that changing the direction of his scientific work approximately every ten years had helped to keep him fresh and productive. He agreed with researchers on creativity who had observed that "a person who commands several branches of knowledge transfers something that is well-known in one area into other areas."

Linus's involvement in the peace movement had given him and Ava Helen a common interest and activity, which was rewarding for them as a couple as well as contributing to a cause. Pauling acknowledged that he had not been an "especially good father," having "left the major part of child upbringing to my wife," but he remained close to his children and grandchildren in his later years.[2] After his wife died, he became much closer to his daughter, Linda Pauling Kamb, who shared many personal matters with him that she would have shared with her mother. He took no interest in other women after Ava Helen died, although after an appearance on the "Donahue" show he received a huge pile of perfumed letters—and one chocolate cake—from women who were eager to meet him. Pauling ate the cake, but he didn't contact any of the women.

His marriage had been a happy and fulfilling one, and despite the women's liberation movement he continued to believe that it was best if a mother remain at home to raise her children. Ava Helen strongly shared his views on this topic. In a letter written to a friend in 1964, referring to Betty Friedan's *The Feminine Mystique,* which sparked a wave of feminist activism in the 1960s, Ava Helen said:

> I think that Betty Friedan and her book do a great disservice to women. . . . Her thesis that any work is preferable to work in the home seems to me to be completely fallacious. . . . I think her scorn and denigration of women and the home and its importance are a disservice to what we should have in the world of the future. . . . The argument that a woman needs something more than her work in the home is, I think, quite valid.

> What I believe is that women and men need to share each other's lives more. And that the reason for the great unhappiness among women, and indeed among men too, is the fact that they have been so separated from any real companionship and any real part in each other's activities.[3]

It is interesting to note that their relationship, which began while Linus was Ava Helen's instructor at Oregon Agricultural College, would have violated sexual harassment codes in many universities today. They avoided problems at the time, Ava Helen said, "because Linus maintained his objectivity in class."[4] In fact, Pauling says that he gave her a grade slightly lower than she deserved, just to be on the safe side. She was clearly the best student in the class.

In a symposium at the Wayne State University School of Medicine in January 1972, Pauling urged the young men in the audience to follow his example:

> I have been especially fortunate for about 50 years in having two memory banks available—whenever I can't remember something I ask my wife, and thus I am able to draw on this auxiliary memory bank. Moreover, there is a second way in which I get ideas. . . . I listen carefully to what my wife says, and in this way I often get a good idea. I recommend to you young people who are participating in this Symposium that you make a permanent acquisition of an auxiliary memory bank that you can become familiar with and draw upon throughout your lives.[5]

Pauling also saw some possible virtue in his financial struggles as a young man, observing that having worked a hundred hours a month in physical labor as a college freshman "may have been profitable" because "it accustomed me to long hours of hard work."

Ever since childhood, Pauling had been an atheist. He found the concept of God problematic and saw no advantage in believing. However, he was not militant about his atheism and had belonged to the Unitarian church of Los Angeles for many years. He and Ava Helen had joined because the church "accepts as members people who believe in trying to make the world a better place."[6] The minister was a leading social activist in Pasadena. Pauling thought of himself as a humanist and took great satisfaction in the belief that he had helped to prevent nuclear war and lessen suffering from cancer and other diseases.

As Pauling's illness progressed and he developed the intestinal cancer that would take his life, he found the administrative work at the institute more of a chore. One day in 1992, his son Linus noticed that paperwork was a burden for his ninety-one-year-old father. Linus Jr. immediately offered to take over and Pauling agreed to make him president and chairman of the board of trustees. At the time, Linus Jr. observed that his father had "not quite as much energy for moving around as he used to have, but his mind is still incredible."[7]

Turning over the administrative chores to his son allowed Pauling to spend more of his time at his home at Big Sur, where he could be alone. There were no neighbors, nearby friends, or other distractions. The days are warm at Big Sur, but the nights are cold. When he was alone at Big Sur, Pauling often accommodated the climate by going to bed at about eight o'clock in the evening. Staying up would have required him to build a fire, which he preferred not to do at night. He typically rose at three or four o'clock in the morning, built a fire, ate breakfast, and settled down to work.

This monastic life-style helped Pauling to maintain his lifelong pattern of extremely high scientific productivity. He estimated that he had published approximately six hundred scientific articles in his lifetime, as well as two hundred papers on social and political topics, several textbooks, and popular books. In his nineties, Pauling continued to do work in molecular chemistry, exploring topics such as the structure of metal hydrides and icosahedral quasi crystals. He sent his papers to the *Proceedings of the National Academy of Sciences,* which accepted them automatically. His name also appeared on coauthored papers in the *Proceedings* on topics such as the suppression of human immunodeficiency virus by ascorbate.

For a number of years, the crusade for vitamin C didn't seem to be getting very far. Pauling "felt rather discouraged, because it seemed that, although our work had progressed well, it was still not accepted by most physicians and in particular by medical authorities."[8] In the early 1990s, however, the tide of opinion seemed to be changing. In 1992, the New York Academy of Sciences held a symposium in Arlington, Virginia, on the value of high doses of vitamins and other nutrients. Participants at the symposium reported on research suggesting that antioxidant vitamins (vitamin C, vitamin E, and beta-carotene) may have a significant effect in combating cancer and heart disease. This theory was bolstered by the findings of several large epidemiological studies.

The new theory was that these vitamins destroyed free radicals (molecules with an odd number of electrons) in the bloodstream. This theory vindicated Pauling's research on vitamins and health over the years, which had had an important scientific impact as well as influencing public opinion. Although the scientists were inspired by Pauling, they often avoided citing him because of the stigma of unconventionality associated with his work on vitamin C. They feared that citing him might make it more difficult to get research funding.

Pauling was unable to attend the conference, but in reading its published proceedings, he came across a comment from Carlos Krumdieck, M.D., a professor of nutrition sciences at the University of Alabama:

> For three days I have been listening to talks about the value of large intakes of vitamin C and other natural substances, and I have not heard a single mention of the name Linus Pauling. Has not the time come when we should admit that Pauling was right all along?[9]

Later, Pauling received a letter from Dr. Krumdieck, telling him that when he made this observation, "The entire audience signaled its agreement by bursting into a loud and enthusiastic round of applause. Though long in coming, I believe that the nutrition community is now prepared to support your views."

The flow of research findings was not, however, universally supportive of Pauling's theories. In April 1994, the *New England Journal of Medicine* published the results of a double-blind experimental study by researchers in Finland who studied a sample of 29,133 male smokers between fifty and sixty-nine years of age.[10] They found that smokers receiving vitamin E or beta-carotene, or both, for five to eight years were no less likely to contract lung cancer than those receiving a placebo. In fact, they were surprised to find that those who received beta-carotene actually had worse outcomes than those who did not.

The Finnish study was so large and rigorous, and the researchers so clearly impartial, that its validity could not be seriously challenged. It was, however, only one experiment with a particular population, and interpreting the findings was difficult. The researchers themselves were quite surprised by the fact that their results contradicted the findings of large, well-conducted epidemiological studies. They speculated that the input of beta-carotene as reported in epidemiological studies might be simply an indicator of other life-style differences, or that there might be other components of fresh fruits and vegetables that accounted for their apparent effectiveness in lessening the risk of cancer. They had no explanation for the higher incidence of lung cancer among those taking beta-carotene and suspected that this finding may have been nothing more than a chance fluctuation. Another clinical trial in China had had the opposite result.

The Linus Pauling Institute was deluged with phone calls and letters asking about this study, which was widely reported. Many supporters wondered whether it disproved claims the institute had been making for many years. In response, the institute issued a statement drawing attention to possible weaknesses in the Finnish study, including the fact that it lasted only eight years and the possibility that smoking may destroy vitamin C, which acts synergistically with vitamin E. They cited the many epidemiological studies and referred to data suggesting that Finnish people may be especially susceptible to cancer.[11]

The issue of vitamins and health continues to be controversial. On a practical level, the most incontrovertible medical recommendation is simply to follow a healthy life-style and eat plenty of fresh fruits and vegetables. There is no compelling experimental evidence that megadoses of water-soluble vitamins are useful. Some epidemiological studies suggest that people who take lots of vitamins live longer and healthier lives, but this may be because these people also do other things that are beneficial.

Pauling's philosophy had always been to have a lot of ideas and then to throw out the bad ones. In his work in the hard sciences, this had worked out quite well. When he made a mistake, such as his embarrassing flub in the race to find the molecular structure of DNA, he accepted the evidence graciously. Advancing speculative hypotheses is a normal part of the scientific method, and Pauling had plenty of successes to balance his occasional failure.

When Pauling moved into the realm of political activism, the rules were different. In political controversy, it is rare that one has a definitive test of one's arguments. Pauling's claim that atmospheric nuclear testing was harmful was proved correct, but there is no way to be certain about what the geopolitical consequences would have been if the United States had followed a different military strategy.

Pauling remained true to his dovish convictions to the end. On January 9, 1991, he published at his own expense an advertisement in the *New York Times* opposing the "rush to war" with Iraq. He argued:

> We are now rushing toward a war in the Middle East, a terrible war that would be fought with highly destructive bombs, poison gas, lethal bacteria, and perhaps nuclear weapons, liberating deadly radioactive fallout over the whole world. It would kill millions or perhaps tens of millions of people. No situation, no dispute, can justify such a war.
>
> We, as individual human beings, must take action to avert the impending catastrophe and say to the leaders of the nations: STOP THE RUSH TO WAR!

This advertisement expressed humanistic concerns shared by a great many people. Pauling believed that American public opinion would turn against the war once American casualties topped the fifty thousand mark as they had in Vietnam and recommended that a negotiated settlement could prevent these losses.[12] The outcome of the Gulf War, however, shows that Pauling's estimates of the costs of war were exaggerated. There is no way of knowing whether the economic sanctions and negotiations he urged would have been more successful. One cannot test the idea with an experiment as one would in chemistry.

Since these political disputes are so ambiguous, most people with strong beliefs remain true to their convictions regardless of how events turn out.[13] Linus Pauling was no exception. His work in orthomolecular medicine followed a similar pattern, even though in this case a more objective, scientific approach might have been more appropriate for a person with his training and abilities. Medical research is often slower than research in chemistry, but there are clear-cut standards that are accepted by reputable researchers. Pauling understood these standards and used them effectively in his own publications, some of which involved complex statistical designs and extensive literature reviews. But when the results of other people's research contradicted his views, he sometimes retreated to the looser standards of the "alternative medicine" community, where testimonials and clinical impressions are often given more weight than results of rigorous experimental findings.

In the last year of Pauling's life, the Linus Pauling Institute of Science and Medicine was plagued with difficulties. In the scientific community it was often perceived as advocating a cause, not doing objective research. Dr. Charles Hennekens, of Harvard University, expressed a common view:

> The Pauling Institute, unfortunately, is perceived as putting forth the research that shows the benefits of vitamin C instead of testing whether or not vitamin C really is beneficial. It's unfortunate (Pauling) became so convinced of the value (of vitamin C) in advance of the data.[14]

When interviewed by a reporter after Pauling's death, Dr. Daniel Steinberg, a vitamin researcher at the University of California, said that he couldn't "think of any significant contributions from the (Pauling) Institute in the last decade." One consequence of this image in the professional community was that the Pauling Institute had great difficulty getting government grants. More than 90 percent of its budget of only $2.4 million a year came from individual contributors.

Even worse, Arthur Robinson was joined in his attacks on the institute by a researcher, Matthias Rath, who filed suit in March 1994, claiming that Pauling had taken credit for his ideas on vitamin C and heart disease. In 1993 Rath, who coauthored several papers with Pauling in the *Proceedings of the National Academy of Sciences,* had self-published *Eradicating Heart Disease,* which argued that heart attacks and strokes were caused by vitamin deficiencies. Experts dismissed the idea as nonsense. The vitamin researcher Daniel Steinberg of the University of California thought that Pauling and Rath's ideas "made little sense," while Ronald Krauss of the Lawrence Berkeley Laboratory said that the book displayed "a strong element of irresponsibility."

Pauling was not concerned about this reaction from established scientists; in fact, he had anticipated it. Rath and he had published an article with the audacious title "A Unified Theory of Human Cardiovascular Disease Leading the Way to the Abolition of This Disease as a Cause of Human Mortality" in the *Journal of Orthomolecular Medicine.* They began the article with a well-known quote from Max Planck: "An important scientific innovation rarely makes its way by gradually winning over and converting its opponents. What does happen is that its opponents gradually die out and that the growing generation is familiar with the idea from the beginning."

Rath and Pauling's new theory was that heart disease is caused by vitamin C deficiency, which causes weaknesses in the vascular walls. In their view, all the problems that heart specialists conventionally treat are merely defense mechanisms that evolved because of the human body's lack of ascorbate. If people would simply take large doses of vitamin C, these problems should clear up. They realized that this was not quite a proven theory, but thought that "the available epidemiological and clinical evidence is reasonably convincing. Further clinical confirmation of this theory should lead to the abolition of cardiovascular disease as a cause of human mortality."[15]

Rath was apparently afraid that Pauling was going to capitalize commercially on his ideas. Throughout the years, Pauling had never endorsed any commercial products, although his books were often sold in health food stores next to the displays of vitamins. In his books, he urged readers to buy the cheapest vitamin C they could instead of paying high prices for "natural" products. The May 1994 issue of *Vitamin Retailer* magazine, however, included the following item:

> A new line of dietary supplements endorsed by Linus Pauling and his family will soon be introduced by Spectrum 2000 of Lynwood. . . . The products, which will feature a picture of Linus Pauling on the label, will be sold through a multilevel marketing system, although possibly also at health food stores.

This marketing scheme, which has been implemented, is reminiscent of Herman Pauling's marketing of "Dr. Pfunder's Oregon Blood Purifier" in 1906. The institute was clearly concerned that this might lead to unfavorable publicity, and a spokesman stated that it was a private project of the Pauling family. Pauling was in the terminal stages of intestinal cancer at the time, and it may be that he went along with this project at the suggestion of some of his children.

Matthias Rath released tapes of a July 23, 1992, meeting of the board of the Linus Pauling Institute at which the members were concerned about Rath's unwillingness to promise to turn over to the institute 75 percent of the profits from any patents that might result from his research at the institute. On the tape, Pauling can be heard commenting, "If it turns out to be a valuable patent, the Institute would get a lot of money."

Pauling's death of cancer on August 19, 1994, ended his personal involvement. Matthias Rath hired the celebrated attorney F. Lee Bailey to threaten his lawsuit against the institute, and the institute reached a settlement.

There can, of course, be no patent on the idea of taking large doses of vitamin C, and the institute continued to distribute Pauling's recommendations freely to anyone who asked. For healthy people, Pauling recommended 6 to 18 grams (6,000 to 18,000 milligrams) of vitamin C every day, together with at least 400 international units (IU) of vitamin E, a B-complex tablet, 25,000 IU of vitamin A or 15 milligrams of beta-carotene, and a multiple vitamin–mineral supplement every day. For good measure, he also recommended minimizing sugar intake, getting moderate exercise, not smoking cigarettes, and avoiding stress.

Pauling followed his own dietary recommendations, taking a very large dose of vitamin C with his breakfast. His son, Linus Jr., a medical doctor, noticed that his father moved his bowels within a half-hour of taking his morning vitamins. He tried to persuade his father to take his vitamin C in smaller doses throughout the day to avoid passing the vitamins through his system. Pauling also enjoyed a hearty breakfast with eggs and breakfast meats, and his institute's dietary recommendations include a statement that meat and eggs are good foods if eaten in moderation. Perhaps the most dangerous of his dietary ideas is the suggestion that a high-fat, low-fiber diet can be compensated for with high doses of vitamin C.

The city of Palo Alto wanted to use the institute's site for public housing. The institute's administrators failed to find a home at Stanford or Caltech and were negotiating with Oregon State University. Oregon State had accepted Pauling's papers, agreeing, as Caltech would not, to his wish that Ava Helen's papers also be archived.

Arthur Robinson was still trying to recover his papers and computer tapes from the 1970s. Early in 1994, the institute had promised to turn them over, but they apparently could not find them. Robinson told the *San Francisco Chronicle and Examiner,* "If those guys have destroyed that much data, I think they should close their doors—I really do. It's a major scandal."

Pauling's obituary in the *New York Times* was suitably laudatory, stressing his work as a "chemist and voice for peace" and briefly dismissing his work on vitamins. On January 1, 1995, the *New York Times Magazine* listed him first in the story "Lives Well Lived." If he had retired gracefully at sixty-five or seventy, his contributions as a scientist and senior statesman of the peace movement would have been unsullied by the controversy of his advocacy for orthomolecular medicine and the conflicts with Arthur Robinson and Matthias Rath. But it would be a mistake to attribute the controversies of Pauling's later years to senility. He remained alert, highly intelligent, and in command of his faculties throughout his life span. At most, one can say that certain traits that were always part of his makeup became more dominant in his later years.

When we review Pauling's life as a whole, we find a number of traits rooted in his childhood that persisted throughout his life. He was an intellectually gifted child who had limited intimacy with his family, especially after his father's death and his mother's long illness. He found nurturance in intellectual life, turning to reading, puzzles, daydreams, and fantasies for entertainment and escape; and he received tangible recognition for his intellectual accomplishments in the forms of scholarships and other academic recognition. Early on he developed a deep inner strength and a tremendous self-confidence, qualities that served him well throughout his scientific career. He learned to channel his aggressive drives into his intellectual life, valuing logic, rationality, and emotional control, and he found joy in his intellectual work.[16]

The major crisis in Pauling's childhood occurred with his father's death when Linus was nine, followed by his mother's long struggle with pernicious anemia. His mother pressured him to sacrifice his scientific aspirations to her and his sister's needs; in response he learned to be determined and self-reliant. Placing his own goals first and foremost, he willfully ignored the norms of his family and neighborhood and distanced himself from his mother's emotional demands. In this respect, Pauling's childhood differed from that of most scientists, whose parental role models were much more supportive of their intellectual inclinations. Pauling's family background required him to find the inner strength to overcome his doubts and anxieties, and resolutely maintain control of his life. He married a woman who eagerly devoted herself to him and their children, and who accepted his scientific goals and shared his political concerns.

Pauling achieved enormous and rapid success through his intelligence, his creativity, and his willingness to shift areas of study in order to follow the most interesting problems. Defying the stereotypic image of the reclusive scientist, he was a brilliant public speaker and greatly enjoyed public attention. He traveled incessantly, finding acclaim everywhere. He was received warmly all over the world, which gave him less time for close personal friendships but was nonetheless deeply satisfying. He lived life on his own terms.

In his early career he set himself apart with his facility for combining chemical and physical ideas. His early work on crystal structure is a classic example of geometrical intuition applied to physicochemical structure, one of the most beautiful such examples in the history of science. His early X-ray crystallographic work contributed substantially to the empirical knowledge base of physical chemistry. But rather than merely continuing to work along these lines, he took a huge step in a new direction with his papers on the chemical bond: instead of geometric intuition, he was applying something much less tangible and more mysterious, a kind of combined quantum mechanical and chemical intuition that seems to have been inaccessible to any other scientist at the time. In one bold stroke, the conceptual gap between the new physics and the old chemistry was bridged. In time Pauling's valence bond approximations were largely superseded by more technical molecular orbital methods, but it was Pauling's ideas that first established a detailed connection between chemistry and quantum physics, thus paving the way for more sophisticated theories.

He continued to study the chemical bond throughout his life, but before long his attention turned to new frontiers. Having resolved to his satisfaction the boundary between chemistry and physics, he turned to the boundary between chemistry and biology—that is, chemistry and life. More than any other single researcher, Pauling laid the foundations for the new science of molecular biology. Many of the basic ideas that molecular biologists today take for granted—helical structure, aperiodic coding of information, and so forth—first emerged in the penetrating work of Linus Pauling's middle years. The successful unraveling of the DNA molecule would have been a natural conclusion to his work on molecular biology, as well as an eminently suitable capstone for his career. However, this was not to be, and thus it may be said that Pauling's career, like those of many scientists, peaked early.[17]

While he continued to receive recognition and awards throughout his middle age, they were more often for past work and less frequently for current accomplishments. This does not in any way detract from his outstanding research of the 1940s and 1950s, which was science at the highest level. However, there is some evidence that this failure to repeat his achievement with the chemical bond theory was a source of frustration for Pauling. While two Nobel prizes may seem to be enough for any man, finding the structure of DNA could have won him a third, and this may have rankled him to a degree. Arthur Robinson told us that Pauling asked him to organize a discreet campaign behind the scenes to nominate him for a third Nobel award—in medicine.

During his years of greatest scientific productivity, Pauling's life was entirely focused on science. His wife managed every detail of his daily life, enabling him to function in an unusually directed, single-minded fashion. As he grew older, however, other parts of his nature began to demand outlets, and he devoted large amounts of energy to activities other than scientific research. Deeply compassionate and courageous, and also capable of intense anger and outrage, he considered it his duty to do what he could to help his fellow human beings avert the threat of worldwide destruction. In this arena, as in his scientific work, he achieved great success through the determined application of his native abilities. Once again, he was speaking to large, enthusiastic audiences. The Senate Internal Security Committee and the right-wing press that tried to stop him were powerful and dangerous adversaries, and Pauling showed rare bravery in confronting them at a time when so many concealed their true colors. In this case, as so often in his life, Pauling triumphed by sticking to his own convictions.

Psychologically speaking, his experience in the peace movement was similar to to his involvement with orthomolecular medicine. The nuclear issue died down after the agreement banning atmospheric testing, and, although there was great satisfaction in seeing the McCarthyists defeated, Pauling missed the excitement of controversy and celebrity. By this time, he had developed a strong belief that the world was polarized between avaricious elites and the deserving common people who needed the support and protection of well-meaning people such as he. When the vitamin C issue came to his attention through a random encounter, he quickly fitted it into this conceptual framework. He believed that the medical establishment was suppressing information that the people needed to protect their health. He did not believe that this issue needed to wait for new research. The truth was clear to him; all he needed to do was publicize it.

The more his views were criticized, the more insistent he became, and the more convinced he became of his convictions. He seemed to regard any criticism or disagreement as an assault on his personal dignity. In his early career, he had learned to stick with his ideas regardless of social pressures, and in most cases his controversial ideas had been proven correct in relatively short order. In the case of orthomolecular medicine, however, his intuitions were not so quickly vindicated. Pauling's advocacy played a useful role in promoting research, and in the end many of his views may be shown to be correct. But his fame in the area of orthomolecular medicine rested less on his intellectual accomplishments than on his eloquence and persistence in advocating a cause supported only by ambiguous evidence.

In the three chief public roles of his life—scientist, political activist, and medical advocate—Pauling consistently displayed an immense single-mindedness and self-confidence. These traits date back to his childhood, and they lie at the root of his brilliant success as a chemist. His deep self-esteem enabled him to put aside any inner doubts, trust his intuitions, and focus his tremendous intellectual resources on his scientific goals. At times these traits may have worked against him scientifically—for instance, he was so pleased with the success of his valence bond theory that he remained committed to that approach throughout his life, even when most chemists had moved on to other theories. But he always welcomed the discipline and challenge of the scientific method, and conceded graciously on those rare occasions when his hypotheses proved wrong. His physicochemical intuition was so good that, in matters of research, his intense faith in himself only infrequently led him astray.

The same self-confidence served Pauling well as an antiwar activist and advocate for free speech during the McCarthy era. The ideological beliefs he adopted during the 1934 Upton Sinclair campaign provided a political framework he would rely on for the rest of his life. In politics, there was no need to submit his ideas to scientific testing. The test that mattered was political effectiveness, and by this standard Pauling did very well, using his scientific prestige and his outstanding skills as a speaker to present his ideas in a persuasive manner. He was able to use his scientific skills effectively when they were relevant, especially with regard to nuclear testing. The only blemish on his political record was a proclivity for using lawsuits too vigorously, not only to defend himself but also to intimidate his critics.

In his third great undertaking, the advocacy of multivitamins and orthomolecular medicine, the results of his intense self-confidence were not so overwhelmingly positive. Pauling's headstrong nature and his faith in his own intuition led him to jump into public advocacy, despite the opposition of the majority of scientists who had worked in the field for many years. He was effective in raising public awareness of the issue and in stimulating new research on a topic that most scientists had abandoned. But when responsible medical researchers disagreed with him, he sometimes took it as a personal assault and evidence of a sinister conspiracy. The bitter conflict with Arthur Robinson, his closest associate and heir apparent to his institute, was not entirely Pauling's fault. It was a personality conflict between two men. But Pauling's oversensitivity to criticism and difficulty in accepting disagreement with his ideas would not allow him to resolve the conflict constructively.

The jury is still out on the role of increased amounts of certain vitamins in human health. However, the Linus Pauling Institute has continued to be burdened with an image as an institution dedicated to proving a point rather than conducting objective research. And, sadly,

Pauling ended his days under legal assault by his own collaborator for questionable nutritional marketing schemes.

Pauling the chemist will go down in history as one of the most remarkable intuitive minds of all time. His breadth of achievement and his ability to propose simple solutions for complex problems have not been equalled by any modern-day scientist. Linus Pauling the man was more than Pauling the chemist and, like all human beings, had flaws as well as virtues. Despite his intensely analytical mind, he was capable of being relentlessly driven by irrational emotions, a fact that became particularly obvious in the last decades of his life. In the long run, Linus Pauling the chemist will be remembered and Linus Pauling the man largely forgotten, but it was the man who made the chemist possible. It was Pauling's stubborn determination and faith in his own intuition that led him to cling to his views on orthomolecular medicine with such excessive tenacity, but it was these same qualities that led him to crusade so courageously for world peace. Most important, it was these same qualities that also gave him the strength to parlay his keen intelligence and creativity into so many outstanding scientific discoveries. In his work on crystal structure, chemical bonding, and molecular biology, his very human personality quirks came together with his almost superhuman intellectual abilities, and the result was wonderful science.

APPENDIX

Patterns in Ink

Linus Pauling participated in two psychological studies of the personalities of scientists: Anne Roe's *The Making of a Scientist* (1953), and Bernice Eiduson's *Scientists: Their Psychological World* (1962). Both studies relied in part on the Rorschach method of personality analysis, a psychological instrument first publicized by Hermann Rorschach in 1921. In the Rorschach, subjects view a series of black and white and colored ink blots and tell the examiner what they think the ink blots depict. The subject examines the ink blots one at a time, and the examiner writes down what the subject says and then goes through the record with the respondent to determine where on the blot each image was seen. Since the ink blots are actually only blobs of ink, the subject must draw on his or her own mind to find anything there.

Psychologists who use the Rorschach test interpret the answers as revealing a great deal about the subject's psychological makeup. There are, however, a range of opinions about the validity and usefulness of the Rorschach test. Some professionals believe that it is best regarded as a clinical tool that can help a therapist in working with a patient, not as an objective measure of personality. Others believe that the Rorschach has some validity as a measure of what is going on in the subject's unconscious mind. These psychologists have developed a number of objective scoring protocols and even computer programs for analyzing responses to the ink blots.[1] They have tested these scoring systems against thousands of cases and found correlations between answers to the test and specific constructs of personality. But even those psychologists who believe the Rorschach has a degree of objective validity recognize that the interpretation depends on the theoretical perspective of the interpreter. All agree that it should be used together with clinical interviews and other diagnostic instruments, not alone.

When Victor and Mildred Goertzel began work on this biography in 1962, Pauling told them that he had taken psychological tests for previous researchers and suggested that these tests might be useful to them. Victor is a psychologist who had used the Rorschach in his Ph.D. dissertation, an experience that left him with a healthy skepticism about the measure. Pauling wrote to Anne Roe, authorizing her to release the Rorschach protocol and she did.

Pauling was an enthusiastic participant in the Rorschach testing. He went through the ten cards in twenty-nine minutes, identifying images as quickly as Anne Roe could record them. At the end of the protocol, she observed:

> Whew! After the first card and his question I did not actively interrupt him but when he came to a pause I picked up another card. He usually but not always put down the one he was holding and took the other, although he could always have gone on almost indefinitely, I don't think he was more hampered at one time than at another. Possibly there would have been fewer on X [the last card] if it had not been apparent that there were no other cards. He quite enjoyed this.

The key images Pauling found in each ink blot are given in the lists that follow.[2]

Card I.

1. pelvis
2. insect . . . like a specimen
3. two pairs of white dots . . . symmetrical translation

4. sine curve
5. lobster claws
6. bat wing . . . I looked for the little hooks a bat uses to hang by but they are not visible.
7. lack of symmetry . . . little white line on the left is not there on the right . . . a little claw there

Card II.

1. blood and the black of ink, carbon and the structure of graphite . . . straight lines in the little central figure are puzzling
2. vulva
3. pair of butterflies . . . wings vertical . . . facing each other
4. pair of sharp-nosed pliers
5. two rabbits . . . in an attitude of supplication

Card III.

1. two men perhaps waiters . . . formal dress . . . facing each other . . . Joos dancers or some other pair of male dancers
2. crab (the men are holding)
3. Picasso . . . two white spots . . . two eyes looking out . . . the nose . . . oligocephalic
4. red blotches . . . the Bible is standing open

Card IV.

1. a pelt skinned off . . . on the skin side and to some extent on the fur side
2. Dali's watches . . . the two arms . . . hang over in that limp manner
3. spigot that iron comes out of a cupola
4. testicles and penis . . . pile of skins (referring to 1)
5. gorilla . . . standing there, illuminated by a bright light close behind his back
6. carcass of an animal spread open; I seem to see a cleaver, not in the picture but the act of cleaving
7. little group of very small dots . . . spots on a Laue photograph . . . two-dimensional lattice

Card V.

1. batty
2. swallow tailed butterfly . . . moth
3. deer . . . horns of a deer in the velvet
4. nut cracker
5. man with a derby hat just below the horns which suggests he is cuckolded
6. Icarus . . . like DaVinci's drawing . . . wearing skis
7. alligator, the heads . . . bulging above the eyes

Card VI.

1. totem pole effect
2. same sort of skin as before
3. the question of embryological development that arises from the ridge down the middle
4. this should be colored and should be orange, I don't know why

Card VII.

1. insect . . . the antennae or some mouth parts
2. animal faces and heads, like the funny papers
3. hinge . . . special sort of structure . . . bivalve
4. crustaceans or lobster claws
5. appearance of islands from the air, but the symmetry tends to remove that because no tropical island would occur in pairs like that

Card VIII.

1. nice colors . . . sort of skeletal, too
2. couple of animals . . . not exactly beaver like, tails to the bottom, climbing up . . . Dutch painter, Breughel? . . . and of Bosch . . . fanciful animals . . . the temptation of St. Anthony involved trumpets in the noses and in this case . . . tail suggests an adhesive organ, like the placenta
3. the color . . . a liver a spinal column of a fish and ribs coming out (refers to 1)

4. one of those Breughel imaginary animals
5. landscape, there has been a lot of erosion by the rain

Card IX.

1. that's Punch, two Punch's . . . with pendulous abdomens
2. insects
3. pelvic bones . . . from in front instead of above (referring to Card I)
4. water is dripping, perhaps blood dripping down
5. peaches or similar fruit, four of them arranged in a row
6. flame produced from a central structure (two elaborations of 1 and 3)
7. holes . . . holes of the metal cylinder into which the glass globe of a kerosene lamp would fit and the bottom structure might be the container for the lamp
8. two pigs heads . . . end of snout a porcine indication

Card X.

1. wish bone
2. governor of a locomotive the jowls . . . 3 ellipses attached together by arms . . . dynamically unsatisfactory
3. facing gnomes, two on the right and two on the left, the fatter one with arms around the thinner holding up a green structure which isn't heavy
4. two similar gnomes holding up, perhaps a candle stick . . . some little insect, colorless, water nymph
5. pelvis
6. a rabbit being held up by
7. two caterpillars
8. nice yellow sea shells, not exactly conch shells . . . some sea shells are spiny
9. sea horses, but the tails are bent the wrong way
10. Irish appearance too, the nose, and there is something hanging from both upper and lower lips, mouth open, it's ectoplasmic
11. the California peninsula, geographical coastal contour
12. the floats that hold kelp upon the surface

13. a sweet pea, not quite open
14. Madagascar
15. locust
16. a cow lying down

To the nonprofessional who has never studied the Rorschach, Pauling's answers seem very imaginative and creative. As one might expect, they use some scientific terminology. There are more references to animals, plants, and geography than to molecular structures. Pauling's lifelong hobby of reading encyclopedias had apparently given him a tremendous wealth of images to draw upon, and he enjoyed the creative process that the test demanded.

Rorschach experts, however, can find a great deal more meaning in these responses than lay people. Ted Goertzel asked his colleague, the psychologist Michael Wogan, to review the Rorschach protocol. Wogan knew that it was Linus Pauling's protocol and took his knowledge of Pauling into account in his interpretations. His interpretation highlighted a number of aspects of Pauling's own personality. Wogan thought that Pauling:

- was extremely ambitious
- used a great deal of effort to protect himself against showing emotion
- tended to establish intellectual distance between himself and others, treating himself and others as objects
- felt considerable emptiness due to the psychic effort devoted to his defenses
- had a pervasive fearfulness, visualizing the world as being crushed, cleaved apart, split, or bloodied
- felt a constant need to be in control, which could make problems in intimate relationships.

Wogan thought that Pauling's marriage was probably one-sided, and that he was generally sexist with women although bright enough to avoid expressing this openly.

The most outstanding feature of Pauling's Rorschach result, in

Wogan's view, was the lack of emotion. Wogan thought that Pauling was a person who felt few of life's pains and pleasures, avoiding strong emotion through denial and defenses.

In order to check on the reliability of the Rorschach interpretation, we compiled a list of twenty-two specialists who had published articles on Rorschach interpretation in the *Journal of Personality Assessment.* We wrote to them and asked them for a "blind" interpretation, knowing nothing but the subject's sex and age at the time of testing. Fortunately, seven of these distinguished Rorschach experts generously agreed to participate in this research, purely on a voluntary basis.

When the experts' reports came in, we were pleased to find that they confirmed many of Michael Wogan's impressions. The fact that they were also consistent with each other in many ways increased our belief in the reliability and usefulness of the Rorschach test. On the other hand, we were quite surprised that the experts found the pathological characteristics they did in Pauling's responses, since Pauling had never required treatment for any kind of psychiatric illness.

The first blind Rorschach interpretation we received was from Clifford DeCato of Widener University. Dr. DeCato has practiced and taught Rorschach interpretation for twenty-five years and has published widely on the topic. He became intrigued with what he called the "Mystery Case," spending as much as fifty hours of his time scoring and analyzing the record. He used two different scoring systems, the Perceptanalytic system developed by Z. A. Piotrowski, and the Comprehensive System developed by John Exner, Jr. He provided us with the computer printouts and scoring records for the systems. Dr. DeCato warned us, however, that there were instances in which he had to make "educated guesses" as to aspects of Pauling's responses, since the psychologist who administered the test was not available for questioning. Several of the other experts, also, had told us that it was not always clear from the record which part of the ink blot Pauling was looking at when he made a particular remark. The record of the session, which took place over forty years ago, was not made with the complete rigor and precision expected of Rorschach records in the 1990s.

Dr. DeCato also warned us that "psychopathology may emerge more dramatically" in the Rorschach than in other tests. This was a useful warning, since his interpretation, based on the Comprehensive System, began with this rather ominous quote from the computer printout (The Rorschach Interpretation Assistance Program): "Warning!!—He has many of the characteristics common to people who effect suicide. The possibility of a suicidal preoccupation should be evaluated carefully, and those responsible for his care should be alerted." DeCato went on to note,

> The composite of findings concerning thinking and perceptual inaccuracy suggests a possibility of schizophrenia. . . . He appears to be prone to frequent episodes of depression or emotional turmoil. . . . He processes information hastily and haphazardly. . . . His conception of himself is not well developed and is probably rather distorted. His self image includes many more negative features than should be the case.

In fact, Pauling was certainly not schizophrenic, he had never shown any signs of being suicidal, nor had he needed anyone to be "responsible for his care."

Using the Perceptanalytic Method, Dr. DeCato's observations were much closer to the mark, although still focusing on the negatives in Pauling's makeup. He found that the "Mystery Case" was a person who

> gives the impression of an adult man who is intellectually very bright and has acquired through reading, education, or experience a wide array of information. He attempts to make his adaptation to the world through the use of his intelligence in a rapid-response fashion . . . he is often quick to respond without taking the time to review the situation in depth. He often responds hastily and avoids searching for a more thorough understanding of the whole. The upshot of this cognitive style is that he may often use his intelligence in relatively superficial ways and may make some errors of judgment by forming his opinions too hastily, or at the very least, not engaging his intelligence to the fullest. . . . At times his judgment can become quite unrealistic and disorganized when he is assessing himself or others. . . . He tends to focus on himself and his own feelings more than most people do which along with other features of his protocol suggests a painful sense of distortion in his self, a sense of being insufficient or damaged in some way, along with tendencies to brood on his own emotions.

DeCato further observed:

> A strong trait of ingrained long standing anger expressed as hostility and a trend toward being oppositional and/or stubborn is a prominent feature of his personality. . . . The need for his own space, to be his own master, to do things his own way, not be controlled by authorities, or to have control over his own life and be independent are some of the possibilities singly or in combination. People with this trait can sometimes accomplish outstanding achievement by refusing to give in and by insisting on following their principles or convictions no matter what the cost.

Dr. DeCato further observed, however, "In appropriately structured situations he might be able to use these features of his personality constructively or creatively."

These observations based on the Perceptanalytic analysis fitted Pauling much better than those based on the Comprehensive System, but the Perceptanalytic system also led DeCato to the observation, "Many problems in thinking, logic, and synthesizing across cognitive categories occurred which in terms of both frequency and type of distortion are similar to individuals who have schizophrenia." DeCato concluded that the subject was a challenging case for Rorschach analysis, a bright, intellectualized man who

> struggles constantly with tendencies toward unrealistic perceptions and judgments which he can keep under control in more superficial situations, but which nevertheless are revealed in odd ideas and associations, leaps and breaks in logic, distortions in self and other perceptions, and emotional misjudgments.

The next psychologist to report was James Kleiger of the Meninger Clinic in Topeka, Kansas. Dr. Kleiger has had fifteen years of clinical, teaching, and supervisory experience with the Rorschach and has published several papers on its clinical uses. Dr. Kleiger observed that the subject was erudite and took pride in his intellectual judgments. He added, however,

> Unfortunately, his good natured attempts to amuse himself and impress the examiner with his knowledge and wit are quite strained and reveal a

> desperate effort to manage his confusing world by relying on an ineffective intellectual style. . . . His responses were infused with a language of scientific precision; however, on occasion, he was unable to actually produce a scorable response. Characterologically, one is left with an impression of an individual with narcissistic, obsessional and histrionic traits. . . . There is evidence that this man is working hard to ward off a clinical depression, most likely associated with his underlying sense of narcissistic vulnerability and deterioration. While not actively psychotic, he reveals some signs of idiosyncratic thinking, especially under the impact of his frantic efforts to fend off an unwanted sense of himself as weak and inadequate.

Dr. Kleiger concluded that Pauling's defensive style was generally ineffective and did not fend off feelings of vulnerability or his "nagging sense of cognitive and physical decline." He thought that Pauling showed a tendency to get caught up in emotionally evocative stimuli that would lead one to wonder about a possible hypomanic condition.

The next psychologist to answer was Paul Lerner of Asheville, North Carolina, who has been a leader in Rorschach analysis for many years and has published standard reference works on the subject. Dr. Lerner thought that

> the subject presented as a highly pressured, manicky, very striving, idiosyncratic individual who is markedly self centered. He is intellectually exhibitionistic and pretentious. He used the test more to show off his vast storehouse of information than to merely comply with the task. . . . Prominent in the subject's character makeup are obsessive compulsive and narcissistic features. . . . He is self-centered, self-absorbed, egocentric and highly sensitive as to how he is regarded and treated by others. . . . With respect to his thinking, he was an exceptionally bright individual who at this time is losing it.

Lerner thought that Pauling was depressed and that this depression was related to declining mental powers in middle age. He observed,

> The most prominent affect to appear on his test was depression. . . . Particularly distressing is his sense of being a shell of the person he once was . . . at the time of testing there were test signs to indicate he was sui-

> cidal. While I cannot assess the acuteness of the danger, it would be related to feelings of helplessness and powerlessness and a sense of inability to regain his lost self-esteem.

In summary, Lerner concluded,

> This once high striving, high-powered, exceptionally bright, proud individual is faltering. Despite attempts to cover it over and compensate for it, he is aware of it and feels it. His brain was exceptionally important to him. It was active, big and powerful, and a source of self-esteem. His pride and joy if you will. It was also his competitive weapon. It is now a source of shame and embarrassment. . . . He is experiencing considerable pain. There is much depressive affect centered around a loss of self-esteem, inner feelings of emptiness, and a sense of being a mere shadow of what and who he once was.

The next interpretation was that of John E. Exner, Jr., executive director of the Rorschach Workshops and creator of the Comprehensive System for analyzing the Rorschach. He expressed some reservations about the protocol, which was "not well taken and apparently the examiner lost control of the situation." He also thought that having some information about marital status and interpersonal relations might have helped to clarify the "obvious issue of loneliness and/or emotional deprivation" that he observed in the protocol.

Exner thought Pauling conveyed "the impression of a very disorganized individual whose thinking currently is fragmented, impulsive, and often quite chaotic. The characteristics of his disorganized thinking are typical of individuals who are unable to control and direct their thinking effectively." This disorganization, in Exner's view, was a chronic feature of Pauling's personality.

This observation is remarkably inconsistent with the information known about Linus Pauling. If there was one thing Pauling could do, better than almost any other human being, it was organize his thoughts effectively (even if his thoughts, like anyone's, were not always accurate). If Pauling did not organize his responses to the ink blots in the way that most people do, perhaps it was simply because he thought the test did not call for organized, systematic thinking, but for a disorganized "brainstorming" process. Pauling had read the literature on creative thinking, and this literature strongly recommends against imposing structure on the initial phases of a creative process. Pauling's skill in doing this may give some insight into how a highly creative person differs from more typical people.

Exner also felt that Pauling was undergoing some kind of "situationally related stress" that gave him "a marked sense of helplessness regarding his ability to respond effectively to the current circumstances." He thought Pauling was "burdened with some very intense negative feelings which included a marked sense of loneliness and a general pessimistic outlook concerning himself and his world." He observed, "It is obvious that he is a very intellectual person and, among other things, is prone to deal with his feelings on a more intellectual level than is customary for most people." This is certainly a valid observation about Pauling, and one that was noted by the other Rorschachers as well.

Exner thought that Pauling

> does not process new information very well even though he makes a very concerted effort to do so. It seems obvious that he had a superior capacity to organize new information, but he often becomes almost obsessively trapped in details and his rather hectic thinking causes him to scan a stimulus field too hastily. . . . This issue of reality testing is complicated even further when issues concerning his self image or self esteem are involved. Under those circumstances, he tends to distort reality considerably. . . . He is not the sort of person who controls his emotional expressions very effectively. . . . He would like to be close to people but feels a marked sense of loss or failure in his attempts to develop close relations with others.

"In summary," says Exner, "it is very likely that this is an individual who will be regarded by those around him as 'crazy.' Certainly, the disorganization of his thinking will convey this impression if one sits and listens to him for any lengthy period of time." In real life, of course, Pauling was a brilliant lecturer who impressed tens of thousands of people with his encyclopedic knowledge, rigorous logic, and brilliant insights.

The next Rorschach interpretation was from Eric Zillmer of Drexel University, who had just completed a book analyzing the Rorschach test results of Nazi war criminals. Dr. Zillmer also noted the deficiencies of the protocol, but thought that it appeared valid in terms of being able to offer meaningful interpretations. He also observed that it was "particularly rich, spontaneous, and included a variety of imagery that would pique the curiosity of any experienced Rorschach analyst." He had it scored separately by two experts, using the Exner Comprehensive System, and the interscorer agreement among all the responses exceeded 80 percent. He then used two different computer programs to generate interpretative hypotheses.[3]

Zillmer thought that Pauling was

> a very bright and capable person who responds inconsistently to new problem solving situations or when making decisions. . . . The protocol further suggests that this individual was experiencing substantial emotional uneasiness or distress at the time of the Rorschach administration. This may be related to a general sense of anxiety and tension, unmet dependency needs, and the internalization of emotional experiences.

Zillmer observed that Pauling

> is somewhat uncomfortable in dealing with emotional experiences or situations directly. . . . Individuals with this style usually feel uncomfortable about their ability to deal with negative feelings adequately and often "bend reality" to avoid dealing with perceived or anticipated negatives in their environment. This may lead to social isolation, a sense of loneliness, or emotional deprivation. This presents a conflict for this subject since there are indications of strong unmet needs for emotional sharing, accessibility, and interpersonal closeness.

Zillmer thought that Pauling had "unusually good internal resources to meet stress demands," but that he "may not be as controlled in situations where there is an increase in confusion about feelings, or when confronted with highly ambiguous situations." He thought that

> a core element in this subject's personality is a narcissistic child-like tendency to overvalue his personal worth. This appears to be a dominant psychological influence which, although not necessarily pathological,

> does have a substantial influence on his perceptions of the world, as well as on decisions and behaviors.

In terms of interpersonal processes, Zillmer observed, "It is likely that this subject tends to be regarded by others as likable and outgoing." He thought that Pauling "tends to demonstrate a substantial flexibility in his cognitive approach to the environment and might be expected to think about the environment in a more varied manner than found among more cognitively rigid and less creative or intelligent individuals." Zillmer thought that Pauling "displays, at a minimum, an unusual response style which neglects the conventional, expected, simple, or acceptable response to his surroundings."

In summary, Zillmer found the protocol to be very unusual:

> most likely given by a highly complex man who has many strengths, but also several liabilities in his personality structure. . . . The present Rorschach inkblot protocol indicates both, the potential for brilliant insight and sophistication on behalf of the respondent, but also the likelihood for inappropriate behaviors ranging from immaturity, to distorted thinking, particularly when confronted with emotionally laden situations. Thus, the central issue which defi.nes the main aspect of the individual's personality structure, is related to how successfully he copes with his affective and emotional world.

Zillmer's interpretations varied considerably from Exner's and DeCato's, despite the fact that he relied, in part, on Exner's computer software, which DeCato also used. This software is based on actuarial data from empirical studies on over forty thousand responses. Exner did not tell us whether he used the software in his own analysis. In checking the computer coding and printouts, however, we found that Zillmer and DeCato had scored the responses in much the same way and received essentially the same computer output. DeCato had stayed closer to the computer output in his interpretation, while Zillmer had used the computer output as a source of hypotheses to be balanced against his overall impression of the personality. On balance, Zillmer's interpretation seemed quite close to Pauling's personality as revealed in the biographical data.

The next Rorschach interpretation is that of Vincent Nunno, a psychologist in Oakland, California, who has also analyzed the Rorschachs of Nazi war criminals. Dr. Nunno stated, "This individual does not show test features which are commonly associated with a diagnosis of mental disorder." However, he thought that Pauling's responses showed a tendency to obsessively break stimuli down into details. He observed,

> It might be "argued" that people are "creative" because they can see reality in "new ways" rather than conventional or consensual ways, and this is probably true, but in this case, it appears that the breakdown is not due so much to "creativity" but rather, to an eccentric, overly unique, pedantic and self-centered style in which the subject does not perceive "conventional" boundaries, but rather, offers loose impressions or creates arbitrary boundaries in an attempt to "create the field" as he wishes to see it or talk about it rather than truly "analyzing" the natural contours and shapes of the blots.

Dr. Nunno observed,

> My own feeling is that this person is using an overly intellectualized, overly self-referenced approach to these blots. He doesn't really "look" at the cards in some "neutral way," trying to figure out what is there. . . . He just assumes he does see reality without questioning and that it all must relate to him and his experience of the world. Possibly this is a characteristic of individuals who are "famous" for their unique intelligence but who have more difficulty with the world as it is "commonly" perceived by the average person.

He also stated, "I am getting the sense that this was once a man with a highly 'functional' intellectual style that is now starting to become less efficient and organized due to aging."

Finally, Dr. Nunno warned that "it is always difficult to evaluate 'creative' or 'exceptional' people with a test that is grounded in the concept or 'normality' as these individuals are not 'normals' in the true sense of the word, and their 'uniqueness' should not be conceptualized as a pathological deviation from normal expectancies."

At the last minute, we received an interpretation from Dr. Richard Kramer, a clinical psychologist in Israel whose busy schedule did not permit him to spend as much time as he would have liked with the record. Dr. Kramer thought that Pauling was

> superficially very bright however his intelligence is more for show rather than what he can actually utilize. . . . He is pedantic and does not think things out in a deep fashion. . . . He is very reactive to his environment and exhibits signs of emotional impulsiveness, defensively, he operates via denial and attempts to psychologically distance himself from things in a narcissistic fashion. The individual is a very aggressive man . . . he has a great deal of hostility and contempt toward women, in this respect there is a great deal of classic masochism. It may even be that he is impotent (however, this is admittedly really pushing it as there is no actual data—this is more inferential).

In real life, of course, Pauling had no problems with women or sexuality, enjoyed life fully with no signs of masochism, and was extraordinarily effective in utilizing his intelligence. When the interpretations diverge so sharply from the reality of Pauling's personality, it is tempting simply to dismiss the Rorschach results as invalid. It would be wrong to reach a conclusion about the validity of the Rorschach as a psychological measure, however, from one case. It may be that there was something in Pauling's personality that made him a particularly difficult subject for Rorschach interpretation. Over the years, many people have observed that genius and madness seem to have something in common. As long ago as 1680, the poet John Dryden wrote:

> *Great wits are sure to madness near allied,*
> *And thin partitions do their bounds divide.*[4]

The Line Between Creativity and Madness

In his biography of Richard Feynman, James Gleick asks, "When people speak of the borderline between genius and madness, why is it so evident what they mean?"[5] Perhaps these Rorschach interpretations, so many of which confuse Pauling's creativity with psychiatric disorder, can help us to answer Gleick's question. There is evidence that highly creative people often score similarly to schizophrenics on the Rorschach, even though they do not have any kind of psychiatric disorder.[6] This is believed to be true because creative people are able to draw on primitive psychological processes that "normal" people do not often use. However, they are not stuck on a primitive or chaotic level of thinking, as some mentally ill people are, but are quite capable of integrating their thinking in a mature way when appropriate.

Bernice Eiduson observed this phenomenon in her study of scientists (in which Pauling was included). She thought that scientists as a group had a heightened sensitivity to experiences that

> is accompanied in thinking by over alertness to relatively unimportant or tangential aspects of problems. It makes them look for and postulate significance in things which customarily would not be singled out. It encourages highly individualized and even autistic ways of thinking. Were this thinking not in the framework of scientific work, it would be considered paranoid. In scientific work, creative thinking demands seeing things not seen previously, or in ways not previously imagined; and this necessitates jumping off from "normal" positions, and taking risks by departing from reality. The difference between the thinking of the paranoid patient and the scientist comes in the latter's ability and willingness to test out his fantasies or grandiose conceptualizations through the systems of checks and balances science has established. . . . One might say that scientific thinking is in a way institutionalized paranoid thinking; it sanctions it not only as proper, but also as the irrational that ultimately promotes the rationality of science.[7]

In many ways, Pauling's personality and life history fit the pattern Eiduson found in many of the scientists she studied. She observed that scientists

- were intellectually gifted children whose greatest talent was their mind
- had limited intimacy with their families as children, particularly with their fathers who were often absent
- found nurturance in intellectual life, turning to reading, puzzles, daydreams, and fantasies for entertainment and escape
- received tangible recognition for their intellectual accomplishments in the forms of scholarships and prizes
- built a set of "intellectual fences" to defend themselves against problems or disturbances at home

- learned to value novelty, innovation, and difference, while tolerating any ambiguity and uncertainty which this might create
- developed into intellectual rebels, channeling their aggressions into their intellectual life
- valued logic, rationality, and emotional control
- were likely to enter into traditional marriages with competent women who took responsibility for home and children
- were fulfilled by their work as an end in itself, not primarily for the extrinsic rewards it provides

These findings fit Linus Pauling remarkably well. In many ways, Pauling's personality profile was much like other distinguished scientists'. Pauling differed from most of the scientists Eiduson studied by his intense involvement in political and medical controversies and his tendency to take very strong positions on issues in which objective evidence was ambivalent at best. To understand these traits, we need to look at the specifics of Pauling's personality structure.

We sent a summary of our results to each of the Rorschach experts, letting them know who the subject was and sending them a copy of the article on Pauling we had published in *Antioch Review* in 1981. In response, Dr. DeCato rose to the defense of the Rorschach, observing, "There are many startlingly consistent points between your *Antioch* article and my Rorschach blind analysis." His "attempt to theorize about what might have occurred with Dr. Pauling" is worth quoting at length:

> The description of Dr. Pauling given in the article by Goertzel, et al., in *The Antioch Review* corresponds on many points with the blind analysis by the Perceptanalytic method. This analysis suggested that he was very bright and intellectualized which means in part that he loved his own ideas more than most things in life. The many instances of unrealistic logic and other instances of "cognitive slippage" which are often found in impaired populations such as schizophrenia were counterbalanced in his personality by a high degree of social awareness and conventionality. In other words, Dr. Pauling was a complex mixture of both conventional trends and highly unconventional trends and at times impulsive and unrealistic thinking.

> The description in the *Antioch* article corroborates this view in many different places. The blind analysis suggested that he would have difficulties with close relationships but might be able to function well in structured environments. Again, the *Antioch* article describes a man who preferred to leave everyday affairs to his wife and subordinates, devoting his time and efforts to creative thinking. The tendency to slip into unrealistic thinking was constant, but was generally countered by his high intelligence and wish to be socially respected. For Dr. Pauling, the conflict probably always existed between believing his own ideas which could be unconventional and his desire to be accepted and respected which is conventional. His tendency to become unrealistic and to believe the reality of his own fantasies over external evidence was both a strength and a weakness. At its best this trend allowed him to be very creative, breaking the usual rubrics and inhibitions of thinking and learned knowledge to produce novel ideas and solutions to problems, a process sometimes referred to as 'regression in the service of the ego.' Indeed, there is every reason to believe that his capacity for sudden breaks in conventional logic may very well have helped him make breakthroughs in his research and scholarly efforts. On the other hand his tendency to detach from reality and violate the usual modes of causal thinking and conventional beliefs very likely contributed to his social problems (the social isolation, arrogance, impulsivity, and ill treatment afforded his colleague, Dr. Robinson). In a word, Dr. Pauling was capable of believing things that others might consider unproven or absurd, and held his own ideas in greater esteem and affection than he did people. To him his ideas were more real and more important than anything else leading him to act in support of his ideas and ignore the emotional and interpersonal consequences.

Dr. DeCato's argument makes a lot of sense, since it selects from the Rorschach interpretation those points that are consistent with Pauling's biography. Psychologists who use the Rorschach with their patients probably do much the same thing, interpreting the results in a way that makes sense given what they know from clinical interviews. Of course, much the same thing could be said of fortune tellers, who can sometimes make impressive interpretations of Tarot cards or other random phenomena by responding to cues from their clients. In the blind analyses, the experts had no way of knowing which of their interpretations was off the mark.

Pauling's Personality: A Biographer's Appraisal

Perhaps the Rorschach can be useful, even when it is unreliable, because it helps us to break out of our established mental sets and confront new hypotheses. In this final section, and in this spirit, we offer our own interpretation of Pauling's personality. This interpretation includes only those points from the Rorschach interpretations we consider consistent with the biographical information. Since Eric Zillmer turned out to be a neighbor as well as an expert in personality assessment, we invited him to review all of the Rorschach interpretations and help us in preparing this appraisal.

There is no question that Pauling was extremely intelligent, having both verbal and mathematical ability. He had an outstanding ability to visualize spatial relationships. He was a creative, intuitive thinker, for whom new ideas came quickly and spontaneously. He contrasted himself to very capable scientists who got new ideas by "fiddling with the equations." By contrast, he said, "I've never made a contribution that I didn't get just by having a new idea. Then I would fiddle with the equations to help support the new idea."[8] His approach, as he often remarked, was to have a lot of ideas and then throw away the bad ones.

He had two different intellectual styles in coping with this flow of ideas. In the first, he carefully tested his ideas against empirical data. In this mode, he was open to modifying or even abandoning his ideas if they were not supported. In this process, he often came up with new ideas. He used this mode of thinking in his work in chemistry, and more generally in work that did not entail a strong emotional dimension. He was at his best when he was solving scientific puzzles. In the second mode of thinking, he became emotionally committed to his ideas and selectively sought out evidence to support them. He became defensive against those who questioned his thinking on these matters, often assuming that they were motivated by personal animosity. He made the strongest case possible for his point of view, while minimizing contradictory evidence. His political and nutritional work often followed this second mode of thinking, and it was often effective in advocating controversial positions.

In contrast to his tremendous enjoyment of intellectual activity, Pauling found emotional life troublesome, and he often tried to avoid situations that entailed emotionally charged interactions. He did this especially when he was young, largely as a way to avoid the demands of his mother and others who wanted to steer him away from his intellectual and scientific interests. Once he achieved success with his theory of the chemical bond, he allowed himself to become involved in issues that were emotionally charged for him. Rather than focusing on personal or family life, however, he felt most comfortable in the public arena, where he could rely on his skill as a speaker and writer and his prestige as a scientist.

A core element of Pauling's personality was a narcissistic tendency to overvalue his personal worth and seek the approval of others for his ideas and accomplishments. He loved giving speeches and receiving the approval of large groups of admirers, and he devoted a great deal of time and energy to travel and public speaking at the expense of his scientific work. His narcissism, while often healthy, was sometimes displayed in an extreme sensitivity to criticism, including a tendency to file lawsuits against his critics.

In his personal life, Pauling was stiff and formal, not the kind of person who enjoyed casual, lighthearted activities. He was happy to leave the responsibility for personal and social matters to his wife, to whom he was quite devoted. He did not spend much time on close friendships that involved meaningful interpersonal commitments. His wife was certain he would never have an affair, because he would not want to spend the time needed to romance a woman. He might have felt isolated or lonely, if it were not for her devoted companionship and the continual stream of attention from admirers around the world.

He had the capacity for brilliant insight, but also for distorted thinking, particularly when confronted with situations that were emotionally laden. In these situations, his intellectual defenses sometimes broke down. The sad confrontation with Arthur Robinson is the most striking example of this pattern. It can also be seen in his response to Dr. Moertel and the *New England Journal of Medicine.*

The personality patterns that Pauling displayed throughout life developed in the period after his father's death. His father had admired him greatly and encouraged his intellectuality. His mother, because of her illness and vulnerability as a widow, was not able to provide the same degree of support. He found that he could use his intellectual brilliance to maintain independence from her and obtain approval from others. He married a woman who gave him the devotion he was unable to get from his mother.

Despite his tremendous success as a young scientist, Linus Pauling was never satisfied. Having won two Nobel Prizes, he felt he deserved a third. When his brilliance as a scientific innovator declined with age, he fell more and more into his second intellectual style. In his later years, his combativeness and defensiveness increasingly triumphed over his brilliance and creativity.

NOTES

Chapter 1

1. Interview with Mildred and Victor Goertzel, March 6, 1965.
2. Interview with John Heilbron, March 27, 1964.
3. "Nova" transcript, June 1, 1977.
4. Robert Paradowski, "The Making of a Scientist," in *Linus Pauling: A Man of Intellect and Action,* ed. Cosmos Japan International (Tokyo: Fumikazu Miyazaki), 81.

Chapter 3

1. Linus Pauling, "Fifty Years of Physical Chemistry in the California Institute of Technology." *Annual Review of Physical Chemistry,* 16 (1965): 4.
2. Friedrich Hund, *The History of Quantum Theory* (New York: Barnes and Noble, 1974).
3. Interview with Victor and Mildred Goertzel, March 5, 1965.
4. Lawrence Bragg, *Atomic Structure of Minerals* (Ithaca: Cornell University Press, 1937), 35.
5. Interview with John L. Heilbron, March 17, 1964, Victor Goertzel present.
6. Ibid.
7. Ibid.
8. "Nova," June 1, 1977.

Chapter 4

1. "Nova," June 1, 1977.
2. George B. Kauffman and Laurie M. Kauffman, "Linus Pauling: Reflections," *American Scientist,* 82 (November 1994): 523.

3. Robert S. Mulliken, *Life of a Scientist* (New York: Springer-Verlag, 1989), 61, 97, 99, 126.
4. Robert Paradowski, "The Structural Chemistry of Linus Pauling," Ph.D. dissertation, University of Wisconsin, 1972, 549 (Ann Arbor: Xerox University Microfilms, 1974).

Chapter 5

1. J. D. Bernal, "The Pattern of Linus Pauling's Work in Relation to Molecular Biology," in *Structural Chemistry and Molecular Biology,* ed. Alexander Rich (San Francisco: W. H. Freeman, 1968), 372.
2. Ibid.
3. Letter from Herman R. Branson, president of Lincoln University, to Dr. and Mrs. Victor Goertzel, May 24, 1984.
4. Letter from Linus Pauling to Ted Goertzel, May 14, 1994.
5. Letter from William Lipscomb to Susan Rabiner, April 26, 1995.
6. Dan Campbell, "Antibody Formation: From Ehrlich to Pauling and Return," in *Structural Chemistry and Molecular Biology,* ed. Alexander Rich (San Francisco: W. H. Freeman, 1968), 166–73.
7. Frank Burnet, *Immunology, Aging, and Cancer* (San Francisco: W. H. Freeman, 1976).
8. Myrna Blath, "Wives of Geniuses," *Cosmopolitan* 171 (August 1971): 84.
9. Mary Terrall, "Interview with Sidney Weinbaum," California Institute of Technology Oral History Project, Caltech Archives, 1986, 53.
10. Ibid., 31.
11. "Interview with Frank Oppenheimer," by Judith R. Goodstein, Caltech Archives, 1985.
12. Interview with Mildred and Victor Goertzel, March 5, 1965.
13. Maxine Block, ed., *Current Biography,* (New York: H. H. Wilson, 1943), 351.
14. Interview with Mildred and Victor Goertzel, August 24, 1965.

Chapter 6

1. Linus Pauling, "A Molecular Theory of General Anesthesia," *Science,* 134 (1961): 15.
2. Linus Pauling, "The Hydrate Microcrystal Theory of General Anesthesia," *Anesthesia and Analgesia* 43 (1964): 1.
3. Linus Pauling, *No More War!* (New York: Dodd Mead, 1958), 68–69, p. 209, p. 208, p. 179, p. 203.
4. Linus Pauling and Daisaku Ikeda, *A Lifelong Quest for Peace: A Dialogue.* (Boston: Jones and Bartlett, 1992), 67.

Chapter 7

1. *The Minority of One,* 3, no. 12 (December 1961), 10.
2. Linus Pauling, "The Genesis of the Concept of Molecular Disease," *Sickle Cell Anemia* 53 suppl. (1972): 1–6.
3. George Beadle, "Recollections," *Annual Review of Biochemistry* 43 (1974) 10–11.
4. Letter, May 12, 1963, Linus and Ava Helen Pauling Papers, Oregon State University.
5. Letter, May 29, 1963, ibid.
6. Letter June 25, 1963, ibid.
7. Letter, June 11, 1963, ibid.
8. Letter to Ava Helen, August 23, 1969.
9. Letter to Linus Pauling, May 7, 1981.
10. "Statement by Linus Pauling about the *St. Louis Globe Democrat,*" October 26, 1960, Linus and Ava Helen Pauling Papers, Oregon State University.
11. *Chicago Tribune,* June 19, 1964, p. 14.
12. Letter to Peter Pauling, June 6, 1965, Linus and Ava Helen Pauling Papers, Oregon State University.
13. "Answer" filed with the Supreme Court of the State of New York, ibid.
14. *National Review,* May 17, 1966, 461, 462.
15. Fred Graham, "Courts Take New Look at Libel Law," *New York Times,* April 24, 1966, 6E.
16. Speech in Oregon State University archives. Reprinted in *Center Magazine* 1, no. 6, September 1968, 66–71.
17. Ibid., 70.

Chapter 8

1. Letter to Clifford Derr, computer file, Linus and Ava Helen Pauling Papers, Oregon State University.
2. Letter to James Higgins of the York, Pennsylvania, *Gazette and Daily.*
3. *Linus Pauling, Vitamin C and the Common Cold* (San Francisco: W. H. Freeman, 1970), 2.
4. Letter on file, Linus and Ava Helen Pauling Papers, Oregon State University.
5. *How to Live Longer and Feel Better* (San Francisco: W. H. Freeman, 1986), 226.
6. Thomas Jukes and Jack Lester King, "Evolutionary Loss of Ascorbic Acid Synthesizing Ability," *Journal of Human Evolution* 4 (1975), 85–88.
7. California Board of Medical Quality Assurance, Decision of the California Board of Medical Quality Assurance in the matter of the Accusation against Michael Gerber, M.D., June 20, 1984, 8.
8. California Board of Medical Quality Assurance, Hearings in the Matter of the Accusations against Michael Gerber, 1458, 1473, 1501.
9. Quoted in Evelleen Richards, *Vitamin C and Cancer* (New York: St. Martin's Press, 1991).

10. Personal communication, March 4, 1985, ibid., 147.
11. Letter of April 29, 1985, ibid., 141.
12. Letter to Salzman, May 4, 1987, ibid., 161.
13. *How to Live Longer,* 234.
14. Deposition of Linus Pauling, vol. 1, Sept. 18, 1980, 157 (see references for details on this and all depositions cited in these notes).
15. Quoted in Jerry Carroll, "The Perils of Pauling," *New West* (October 1979): 39–54.
16. Deposition of Richard Hicks, January 28, 1981, 159.
17. Deposition of Linus Pauling, 7.
18. Carroll, "Perils of Pauling," James Grant, "Of Mice and Men: The Linus Pauling Institute Is Plunged into Controversy," *Barron's* June 11, 1979; Farooq Hussain, "The Linus Pauling Institute: An Investigation," *New Scientist* July 28, 1977.
19. Untitled tape, Oregon State University. Tape was probably erroneously labeled Aug. 15, 1962, but parts of it appear to be from 1979–80.
20. Carroll, "Perils of Pauling," 54.
21. Arthur Robinson, "Letter to the Editor," *Antioch Review* (Summer 1981): 383.
22. Linus Pauling, "Letter to the Editor," *Antioch Review* (Spring 1981): 257.
23. Deposition of Linus Pauling, vol. 1, Sept. 18, 1980, 44–45.
24. Ibid., 62.
25. Ibid., 148.
26. Ibid., 149.
27. Ibid., 221.
28. Ibid., 251.
29. Carroll, "Perils of Pauling," 39–54; Grant, "Of Mice and Men."
30. Robinson, "Letter to the Editor," 384.
31. L. Pauling, R. Willoughby, R. Reynolds, B. E. Blaisdell, and S. Lawson, "Incidence of Squamous Cell Carcinoma in Hairless Mice Irradiated with Ultraviolet Light in Relation to Intake of Ascorbic Acid (Vitamin C) and of D, L-a-Tocopherol Acetate (Vitamin E)," *International Journal for Vitamin and Nutrition Research* September 23, 1981, 53–82.
32. Arthur Robinson, Arnold Hunsberger, and Fred Westall, "Suppression of Squamous Cell Carcinoma in Hairless Mice by Dietary Nutrient Variation," *Mechanisms of Aging and Development* 76 (1994), 201–214.

Chapter 9

1. John Woestendiek, *Philadelphia Inquirer,* June 28, 1992.
2. Linus Pauling and Daisaku Ikeda, *A Lifelong Quest for Peace: A Dialogue* (Boston: Jones and Bartlett, 1992) 18, 10.
3. Letter on computer file, Oregon State University.
4. Robert Paradowski, "Chronology," in *Linus Pauling: A Man of Intellect and Action,* ed. Cosmos Japan International (Tokyo: Fumikazu Miyazaki, 1991), 190.
5. Linus Pauling, "The Genesis of the Concept of Molecular Disease," *Transactions*

of the Twentieth Annual Symposium on Blood (Detroit: Wayne State University School of Medicine, 1972), p. 6.

6. Pauling and Ikeda, *Lifelong Quest,* pp. 6, 22.
7. *San Jose Mercury News* July 31, 1992.
8. *Linus Pauling Institute of Science and Medicine Newsletter* Spring 1993, 1.
9. Ibid.
10. The Alpha-Tocopherol, Beta Carotene Cancer Prevention Study Group, "The Effect of Vitamin E and Beta-Carotene on the Incidence of Lung Cancer and Other Cancers in Male Smokers" *New England Journal of Medicine* 330 (April 15, 1994): 1029–35.
11. Zelek Herman and Stephen Lawson, *Vitamin E, Beta Carotene and Disease in Middle-Aged Finnish Smokers,* (Palo Alto: Linus Pauling Institute of Science and Medicine, 1994).
12. Linus Pauling, "Best Time Is Now," *Los Angeles Times* February 18, 1991, B5.
13. See Ted Goertzel, *Turncoats and True Believers* (Buffalo: Prometheus, 1992), for examples.
14. All references in this section from Kesay Davidson, "Controversy Trails Pauling to Grave," *San Francisco Chronicle and Examiner* August 28, 1994, C1.
15. Matthias Rath and Linus Pauling, "A Unified Theory of Human Cardiovascular Disease Leading the Way to the Abolition of This Disease as a Cause of Human Mortality," *Journal of Orthomolecular Medicine* 7, no. 1, p. 12.
16. In all these respects his personality profile is similar to that of many of the scientists described in Bernice Eiduson, *Scientists: Their Psychological World* (New York: Basic Books, 1962).
17. See Dean Keith Simonton, *Scientific Genius: A Psychology of Science* (New York: Cambridge University Press, 1988).

Appendix

1. Howard Lerner and Paul Lerner, "The Rorschach Inkblot Test," in *Test Critiques,* vol. 4 (Kansas City, Mo.: Test Corporation of America, 1985), 523–52.
2. This is a condensed, summarized version, which does not give the locations of the images on the cards. Psychologists who wish to analyze the protocol for themselves should contact Ted Goertzel at Rutgers University, Camden NJ 08102, or the Linus and Ava Helen Pauling Papers at the Kerr Library, Oregon State University, Corvallis, Oregon 97331, for a complete copy. Copies of the complete Rorschach analyses by the experts will also be deposited with the archives.
3. John Exner, J. B. Cohen, and H. McGuire, *Rorschach Interpretation Assistance Program—version 2* (Asheville N.C.: Rorschach Workshops, 1991). Eric Zillmer and R. P. Archer, *The Rorschach Data Sheet Summary and Interpretive Report for Adults* (Indiatlantic, Fla.: Psychologistics Software, 1985).
4. *Absalom and Achitophel,* pt. I, l. 150.
5. James Gleick, *Genius: Richard Feynman and Modern Physics* (London: Abacus, 1992), 313.

6. Charles Hersch, "The Cognitive Functioning of the Creative Person: A Development Analysis," *Journal of Projective Techniques* 26 (1962): 193–200.
7. Bernice Eiduson, *Scientists: Their Psychological World* (New York: Basic Books, 1962), 107.
8. George Kauffman and Laurie Kauffman, "Linus Pauling: Reflections," *American Scientist* 82 (November 1994): 523.

Selected References

Alexander, Rich, ed. *Structural Chemistry and Molecular Biology.* San Francisco: W. H. Freeman, 1968. This festschrift volume includes a reprint of Pauling's classic article on the chemical bond. It also includes a bibliography of Pauling's scientific publications.

American Psychiatric Association. *Megavitamin and Orthomolecular Therapy in Psychiatry.* Task Force Report 7, July 1973.

Cameron, Ewan, and Linus Pauling. *Cancer and Vitamin C.* Menlo Park, CA: Linus Pauling Institute, 1979.

Carroll, Terry. "The Perils of Pauling." *New West,* October 8, 1979, pp. 39–54.

Davidson, Kesay. "Controversy Trails Pauling to Grave." *San Francisco Chronicle and Examiner,* August 28, 1994, p. C1.

Depositions of Linus Pauling, Arthur Robinson, Emile Zuckerkandl, Richard Hicks, Arthur Cherkin, Alejandro Zaffaroni, Dorothy Munro, Corrine Gorham, and Keene Dimick in the Superior Court of the State of California in and for the County of San Mateo, in the matter of Arthur Robinson vs. The Linus Pauling Institute of Science and Medicine, 1980–81. These depositions will be deposited with the Linus and Ava Helen Pauling Papers, Oregon State University.

Eiduson, Bernice. *Scientists: Their Psychological World.* New York: Basic Books, 1962.

Fried, John. *Vitamin Politics.* Buffalo: Prometheus, 1984.

Garfield, Eugene. "Linus Pauling: An Appreciation of a World Citizen–Scientist and Citation Laureate." *Current Contents* August 21, 1989, 9.

Goertzel, Ted. *Turncoats and True Believers.* Buffalo: Prometheus, 1992.

Goertzel, Ted, Mildred Goertzel, and Victor Goertzel. "Linus Pauling: The Scientist as Crusader." *Antioch Review* (Summer 1980), 371–382. A reply by Linus Pauling appeared in spring 1981 (pp. 256–257) and a rejoinder by Arthur Robinson in summer 1981 (pp. 383–385).

Goodell, Rae. *The Visible Scientists.* Boston: Little, Brown, 1977.

Grant, James. "Of Mice and Men." *Barron's* (June 11, 1979), pp. 14–15.

Hawkins, David, and Linus Pauling. *Orthomolecular Psychiatry: Treatment of Schizophrenia.* San Francisco: W. H. Freeman, 1973.

Hussain, Farooq. "The Linus Pauling Institute: An Investigation." *New Scientist* July 28, 1977.

Linus Pauling Institute of Science and Medicine Newsletter. Various issues.

Mead, Clifford, ed. *The Pauling Catalogue: Ava Helen and Linus Pauling Papers at Oregon State University.* Corvallis: Oregon State University, 1991. Includes "Chronological Biography of Linus Pauling" by Robert Paradowski. Also includes a comprehensive chronological bibliography of Pauling's publications from 1920 to 1990.

Paradowski, Robert. "The Making of a Scientist," in *Linus Pauling: A Man of Intellect and Action.* ed. Cosmos Japan International Tokyo: Fumikazu Miyazaki, 1991. This volume also includes an excellent chronology in English and many photographs. Much of the text is in Japanese.

Paradowski, Robert. "The Structural Chemistry of Linus Pauling." Ph.D. diss., University of Wisconsin, 1972. Available from Ann Arbor: Xerox University Microfilms, 1974.

Pauling, Linus. "Fifty Years of Physical Chemistry at the California Institute of Technology." *Annual Review of Physical Chemistry* 16 (1965): 1–14.

Pauling, Linus. "Fifty Years of Progress in Structural Chemistry and Molecular Biology." *Daedalus* (Fall 1970), pp. 988–1015.

Pauling, Linus. "The Genesis of the Concept of Molecular Disease." *Sickle Cell Disease* 53 suppl. (1972): 1–6. *Transactions of the Twentieth Annual Symposium on Blood* (Detroit: Wayne State University School of Medicine, 1972).

Pauling, Linus. "How My Interest in Proteins Developed." *Protein Science* 2 (1993): 1060–63.

Pauling, Linus. *How to Live Longer and Feel Better.* New York: W. H. Freeman, 1986.

Pauling, Linus. *No More War!* New York: Dodd, Mead, 1962.

Pauling, Linus. "On the Orthomolecular Environment of the Mind: Orthomolecular Theory." *American Journal of Psychiatry* 131 (1974): 1251–57. Followed by comments by Richard Wyatt and Donald Kleen.

Pauling, Linus. "The Theory of Resonance in Chemistry." *Proceedings of the Royal Society* 356 (1977): 433–41.

Pauling, Linus. *Vitamin C and the Common Cold.* San Francisco: W. H. Freeman, 1970.

Pauling, Linus, and Daisaku Ikeda. *A Lifelong Quest for Peace: A Dialogue.* Boston: Jones and Bartlett, 1992.

Richards, Evelleen. "The Politics of Therapeutic Evaluation: The Vitamin C and Cancer Controversy." *Social Studies of Science* 18 (1988): 653–701.

Richards, Evelleen. "Showdown at High Noon for Linus Pauling." *Science in Culture* 2 (1992): 282–95.

Richards, Evelleen. *Vitamin C and Cancer.* New York: St. Martin's Press, 1991. An excellent review of the controversy between Pauling and the *New England Journal of Medicine,* with extensive references.

Roe, Anne. *The Making of a Scientist.* New York: Dodd, Mead, 1953. Linus Pauling is included in this study.

Serafini, Anthony. *Linus Pauling: A Man and His Science.* New York: Paragon, 1989.

White, Florence M. *Linus Pauling: Scientist and Crusader.* New York: Walker and Company, 1980. This young people's book was adapted (with permission) from a draft by Mildred and Victor Goertzel.

Williams, R.J.P. "The Political Scientist." *Nature* 342 (November 1989): 235–36.

Zuckerman, Harriet. *Scientific Elite: Nobel Laureates in the United States.* New York: Free Press, 1977.

INDEX